Supercapacitors for the Next Generation

Edited by Daisuke Tashima
and Aneeya Kumar Samantara

Published in London, United Kingdom

IntechOpen

Supporting open minds since 2005

Supercapacitors for the Next Generation
http://dx.doi.org/10.5772/intechopen.94792
Edited by Daisuke Tashima and Aneeya Kumar Samantara

Contributors
Imtiaz Ahmed Khan, Litty Thekkekara, Sania Waqar, Nauman Choudhry, Sabu John, Abd Elhamid M. Abd Elhamid, Heba Shawkey, Ahmed Asaad Ibrahim A.I. Khalil, Iftitan M. Azzouz, Chandra Sekhar Rout, Aditya Sharma, Olivier Fontaine, Yachao Zhu, Angélica María Baena Moncada, Antony Bazan-Aguilar, Elvis O. López, Miguel Ponce-Vargas, Abu Jahid Jahid Akhtar, Amar Laxman Jadhav, Sharad Laxman Jadhav, Anamika Vitthal Kadam, Derya Bal Altuntaş, Sema Aslan, Thibeorchews Prasankumar, Sujin Jose, Jemini Jose, Sreeja P. Balakrishnan, Santosh J. Uke, Sagar D. Balgude, Satish P. Mardikar, A. Bharathi Sankar Ammaiyappan, Seyezhai Ramalingam

First published in London, United Kingdom, 2022 by IntechOpen
IntechOpen is the global imprint of INTECHOPEN LIMITED, registered in England and Wales, registration number: 11086078, 5 Princes Gate Court, London, SW7 2QJ, United Kingdom
Printed in Croatia

British Library Cataloguing-in-Publication Data
A catalogue record for this book is available from the British Library

Additional hard and PDF copies can be obtained from orders@intechopen.com

Supercapacitors for the Next Generation
Edited by Daisuke Tashima and Aneeya Kumar Samantara
p. cm.
Print ISBN 978-1-83968-322-0
Online ISBN 978-1-83968-323-7
eBook (PDF) ISBN 978-1-83968-569-9

For EU product safety concerns:
IN TECH d.o.o., Prolaz Marije Krucifikse Kozulić 3, 51000 Rijeka, Croatia, info@intechopen.com or visit our website at intechopen.com.

Meet the editors

Daisuke Tashima received his doctoral degree (Ph.D.) in material and energy engineering from the University of Miyazaki, Miyazaki, in 2008. He worked as a postdoctoral researcher in the Faculty of Agriculture, at the University of Miyazaki since April 2008, as an assistant professor of Interdisciplinary Research Organization, at the University of Miyazaki since January 2010, as a researcher of Interdisciplinary Research Organization at the University of Miyazaki since April 2014, as an assistant professor in the Department of Electrical Engineering at the Faculty of Engineering at the Fukuoka Institute of Technology (FIT) since April 2015, as an associate professor in the Department of Electrical Engineering, at the Faculty of Engineering, Fukuoka Institute of Technology since April 2018. Currently he is a professor in the Department of Electrical Engineering, at the Fukuoka Institute of Technology. His research interests include supercapacitors, battery, and metal-air fuel cells.

Dr. Aneeya Kumar Samantara, Ph.D., is presently working as a postdoctoral fellow in the School of Chemical Sciences, National Institute of Science Education and Research (NISER), Khurda, Odisha, India. Additionally, Dr. Samantara is working as an advisory board member on the editorial board of a number of publishing houses. Recently, he is a community board member of Materials Horizons, one of the leading journals in materials science of the Royal Society of Chemistry (RSC), London. He has authored 27 peer-reviewed international journal articles, 15 books (with *Springer Nature, Nova Science, ArclerPress, and IntechOpen*), and eight book chapters. Several publications are in the press now and are expected to be published soon. He pursued his master's degree (MSc) in advanced organic chemistry at the Ravenshaw University, Cuttack, and Ph.D. at CSIR-Institute of Minerals and Materials Technology, Bhubaneswar, Odisha, India. Dr. Samantara's research interests include the synthesis of metal oxide/chalcogenides and graphene composites for electrochemical energy storage and conversion applications.

Contents

Preface

Thank you for choosing this book, which summarizes the latest research on supercapacitors. The concept of energy storage is necessary to provide a stable energy supply, and it will thus continue to develop in the future. Supercapacitors, as one of the many energy-storage technologies that have been developed so far, show the higher performance of electrode materials, electrolytes, and separators and demonstrate high energy density as well. In recent years, there has been noticeable progress in the development of hybrid capacitors with larger capacity compared to conventional electric double-layer-type capacitors. Supercapacitors are thus transforming from auxiliary power sources for lead-acid and lithium-ion batteries to next-generation power sources. Going forward, it is expected that the development of supercapacitors using ions, such as sodium and magnesium, and solid electrolytes will keep up the pace. I believe that it is important for researchers in this field to work together and combine technologies to create better products. I hope that this book will act as a platform for researchers to collaborate with each other, and I look forward to the success of the readers.

Daisuke Tashima
Department of Electrical Engineering,
Faculty of Engineering,
Fukuoka Institute of Technology (FIT),
Fukuoka, Japan

Aneeya Kumar Samantara
Academy for Scientific and Innovative Research,
CSIR-Institute of Minerals and Materials Technology (IMMT),
Bhubaneswar, India

Section 1

Fabrications and Performance Evaluation

Chapter 1

Supercapacitors Fabrication and Performance Evaluation Techniques

Imtiaz Ahmed Khan, Litty Thekkekara, Sania Waqar, Nauman Choudhry and Sabu John

Abstract

Supercapacitors have surfaced as a promising technology to store electrical energy and bridge the gap between a conventional capacitor and a battery. This chapter reviews various fabrication practices deployed in the development of supercapacitor electrodes and devices. A broader insight is given on the numerous electrode fabrication techniques that include a detailed introduction, principles, pros and cons, and their specific applications to provide a holistic view. Key performance parameters of an energy storage device are explained in detail. A further discussion comprises several electrochemical measurement procedures that are used for the supercapacitor performance evaluation. The performance characterization section helps to determine the correct approach that should be utilized for supercapacitor device performance measurement and assessment.

Keywords: supercapacitors, electrodes, nanomaterials, fabrication methods, energy storage, performance evaluation, capacitance, and energy density

1. Introduction

The crude energy resources that power our planet are depleting and have a devastating effect on our habitat. The world has turned into an energy soaking sponge always deficient and in need of more reserves to fuel the energy shortage. Toxic fossil emissions are pounding the already battered environment resulting in global warming [1]. Planet Earth is witnessing a drastic climatic change due to this phenomenon. Abrupt weather patterns and frequent ecological calamities events are a few of the core issues arising due to these factors. Humans have realized renewable energy is the only way forward to preserve their environment for future survival. This evolution from fossil-based to renewable energy requires a long transition time and comes with its own various challenges. When renewables such as wind, sunlight, tidal, geothermal, etc., are used to generate electric energy, it requires an efficient energy storage source to ensure uninterrupted and reliable energy supply to the users. Energy storage applications have evolved to cater to various needs, from electric grid-level storage to powering small wearable on-person devices. Electric batteries, fuel cells, capacitors and supercapacitors (SCs) are vital components of energy conversion and storage systems. Electric double-layered capacitors (EDLCs), ultracapacitors, electrochemical capacitors (ECs), pseudo-capacitors, supercapattery are other names that are used for an SC device depending upon the charge storage mechanism [2–4].

In performance metrics, a supercapacitor falls in between a conventional capacitor and a battery. The advantage that supercapacitor exhibits over other conventional batteries are mainly related to a high specific power, significantly high number of cycle life, charge–discharge efficiency, robust thermal operating window and effective handling of fluctuating input–output energy conditions [1, 5–7]. These aspects are summarized in **Table 1**.

2. Components of a supercapacitor

In fundamental form, components of a supercapacitor consist of two electrodes, an electrolyte and a separator that is identical to a conventional capacitor. Supercapacitors (SCs) are electrochemical capacitors (ECs) [9] that store charge in the electric field of electrochemical double-layer [6, 10, 11]. They are one of the favorable candidates for energy storage because of their exceptional electrochemical properties.

Depending upon the charge-storage mechanism of SCs, they can be classified into three; electric double-layer capacitors (EDLCs), redox electrochemical capacitors (RECs) and hybrid electrochemical capacitors (HECs) [10]. Electrodes of supercapacitors can be produced using various forms of carbon [12–14], metal oxides [14, 15] and conductive polymers [16, 17].

EDLCs work on the principle of energy storage by the charge separation at the electrode/electrolyte interface, and they are majorly focused on the materials based on carbon; activated carbons and graphene, carbon nanotubes [18]. While RECs are based on metal oxides, conductive polymers and doped carbon [19]. HECs combines the properties of the materials mentioned above and their working principles.

2.1 Electrode material

Electrodes are the one of the key components and the most important element in SCs. The electrochemical performance of SCs depends upon the properties of electrode materials used in their development. Plenty of researchers are working on designing low-cost, high-performance electrode materials with high stability, high specific surface area and high electronic conductivity [20]. Carbon-based materials are among the popular electrode materials for SCs, followed by conducting polymers and metal oxides, etc. [21].

	Energy storage devices		
Performance indicators	**Battery**	**Supercapacitor**	**Capacitor**
Specific power (W/Kg)	<1000	500–10,000	$>10^4$
Specific energy (Wh/Kg)	10–100	1–10	<0.1
Charge/discharge efficiency (%)	70–85	85–98	>98
Charging time	1–5 hrs	Sec–min	10^{-6}–10^{-3}
Discharging time	0.3–3 hrs	Sec–min	10^{-6}–10^{-3}
Cycle life	~1000	~500,000	>500,000

Table 1.
Performance comparison of capacitor, supercapacitor and battery. Adapted from [8].

2.1.1 Activated carbon

Carbon can be transformed in various forms with a very high specific surface area because of its highly porous structure. This is one main reason for using carbon as an electrode material. Activated carbons (ACs) have proved applications in energy storage [8]. Carbon is abundant in the environment, and its activation can be done through physical (thermal) and chemical activation. Hot gasses are used to develop the structure into ACs in physical (thermal) activation. Carbonization is usually done at very temperature (~500–1100°C). While chemical activation needs lower temperatures (~400–800°C), its pyrolysis and activation are carried out in the presence of dehydrating agents.

Depending on the hybridization, carbon has different allotropes; graphitic carbon has graphene layers, while non-graphitic carbon lacks the long-range 3-D network. The structural characters of ACs are close to the structural properties of pure graphite. ACs can be produced with different porous structures [22]; micropores (nanopores), mesopores and macropores. These pores are important in the kinetics of adsorption and do not increase the adsorption capacity. Changing the factors during carbonization and activation can lead to ACs with different porous structure areas. Conway et al. [5] reported that large pore sizes result in higher power densities while smaller pore sizes relate to higher energy densities.

2.1.2 Graphene

Graphene is a monolayer of carbon atoms packed into a honeycomb lattice and is theoretically regarded as the basis for the formation of all other sp^2 allotropes of carbon. It has excellent mechanical properties and a large surface area with great electronic transportability and thermal conductivity [23–25]. Different methods such as hummer's method [26], dispersion method [27], microwave method [28] are being used. The chemical vapor deposition (CVD) method etc., can be used for the synthesis of graphene. To enhance the capacitance performance of graphene that is synthesized by the methods mentioned above, researchers switched to doping graphene [29], conductive polymer composites [30] or oxide materials [31] to enhance the electrochemical behavior of the materials.

2.1.3 Graphite oxide and graphene oxide

Graphite oxide is a product of graphite oxidized by oxidants such as acids [32], while graphene oxide (GO) is single or few layers of graphite oxide [33]. It can be obtained when graphite oxide suspension is sonicated or stirred. Its properties can be tailored via functionalization of groups on the surface.

2.1.4 Carbon nanotubes

CNTs are another allotrope of carbon. Graphene that can be rolled at a certain axis to produce SWCNTs [34]. CNTs are chemically and thermally resilient and have the highest strength to weight ratios [35]. CNTs can be produced by various techniques, including arc discharge [36], laser ablation [37]. high-pressure carbon monoxide disproportionation [38] and CVD [39]. CVD is the most common method used to synthesize CNTs. It can be produced as SWNT, DWNT and MWNT [40].

2.1.5 Metal oxides/hydroxides

Metal oxides are considered a very good material for supercapacitors because of their very high capacitance and high power, making them very attractive for commercial applications [3]. MnO_2 [41–43], NiO [44–46], RuO_2 [47–49], Co $(OH)_2$ [50] and MoC_3 [51, 52] have got interest because of their vast application for charge storage in supercapacitors. To be eligible for the use in supercapacitors, they must be conductive and can exist in oxidation states without the phase change. While, during the redox reactions, the protons should freely intercalate in and out of the material's lattice.

2.1.6 Conductive polymers (CPs)

Conducting polymers are being explored for redox electrochemical capacitors [53–55] as they have a reversible and fast oxidation and reduction process during energy storage [3]. Due to their high capacitance and large surface area, conducting polymers are being used in supercapacitors. PANI and PEDOT are the most commonly used conducting polymers in supercapacitors.

2.1.7 Composites materials

Hybrid supercapacitors that involve a combination of carbon materials with metal oxides or conducting polymers utilize composite materials [56]. They incorporate the characteristics of both, double layer of charge and faradaic mechanisms. They display higher capacitance to other electrodes that are based on polypyrroles or CNTs [57, 58].

2.2 Electrolyte

An electrolyte is a chemical compound when dissolved in a solvent and dissociated in ions. These ions provide ionic conductivity between the positive and negative electrodes of the device, thus helps in electric charge transportation. The electrolyte plays a vital role in the supercapacitor performance, life cycle, and safety of the device. Chiefly the electrolytes used in supercapacitors are classified into three types, a) aqueous electrolytes, b) organic electrolytes, and c) ionic liquid. Each class has its distinct features related to voltage window and ionic resistance [59, 60].

2.3 Separator

A separator could be any physical barrier such as filter paper, polymeric microporous sheet or even a gel polymer electrolyte that is present between the two positive and negative electrodes to prevent electrical shorting by physical contact of electrodes. Separators should be an inert element and permeable for the electrolyte ions [61, 62].

3. Fabrication techniques

With the advancement in electrochemical supercapacitor technology, the need to design a scalable, sustainable and cost-effective electrode manufacturing method has developed too. Various techniques are being used for the fabrication of supercapacitor electrodes; every technique has merits and demerits over each other. Few commonly used ones are discussed here.

3.1 Chemical vapor deposition (CVD)

This method has gained the interest of many researchers working in this field. The main component used in this method is hydrocarbon substance, as a source of carbon. Iron, cobalt, nickel are some transition metals mainly used as catalysts. Comparatively, in the CVD method, it is easy to control the reaction process. It requires a low growth temperature and is suitable for the production of CNTs and carbon nanofibers. By optimizing the application of catalysts, this method can also align carbon nanotubes arrays [63, 64] that are used to prepare carbon nanotubes [65, 66].

3.2 Dip-coating

Dip coating is a commonly used technique to create substances. It has been reported in the most recent literature of supercapacitors [67–69]. Dip coating is a process where the substrate is dipped into a solution in the presence of a weighing roller/pressure to form a film or a coating on the surface of the substrate. The method is widely used in various industries and in the textile process [70]. It is one of the key techniques used for dyeing. The technique is highly suitable for nanomaterials for creating a thin film coating, such as bio-ceramic nanoparticles, biosensors, and nanocoated implants. The thickness of the coating material affects the adsorption and absorption of the material. The fabric structure, thickness and volume of liquid also affect its absorbance. The dip-coating process is usually followed by air-drying/curing process. The first textile-based supercapacitor was fabricated using this method [71].

The dip-coating method improves the bond between the fibers and the applied electrochemically active materials [72], thus enhance its mechanical properties. CNT/MnO2/PVA fiber electrode was developed by forming a uniform MnO2-PVA paste [72]. The paste was then used to dip-coat CNT fibers. The developed asymmetric supercapacitor showed a wide operating potential window of 2.0 V with the highest energy and power densities of 42.0 Wh kg^{-1} and 19,250 W kg^{-1}, respectively. Cotton fabric was dip-coated using carbon nanofibers (CNFs) to develop flexible carbon composite electrodes in [73]. In order to enhance the electrochemical performance of the electrodes, further layers of manganese oxide (MnO2) and activated carbon were added. Asymmetric supercapacitors (SCs) were assembled using the textile electrodes, which at low discharge rates, exhibited capacitance performance of 134 and 138 F g^{-1} with Nafion membrane and porous paper, respectively. The stable performance of hybrid textile-based supercapacitors using a simple development approach and low-cost materials suggests the future direction for flexible energy storage applications.

3.3 Electrochemical deposition

This is a technique where the electrons are transferred through anions and cations under the external electric field. To form a plating layer, a redox reaction takes place on an electrode [74]. Electrochemical deposition is extensively used to improve the capacitive properties of fiber electrodes by forming nanostructure crystals or conductive polymers. With highly capacitive active materials, this method simplifies the composition between carbon nanotube-based fiber.

3.4 Inkjet printing

Inkjet printing is considered an important breakthrough in manufacturing energy storage devices, particularly in supercapacitors. Over the other fabrication techniques, inkjet printing technology has various advantages such as controlled material deposition, low cost, and compatibility with a variety of substrates [75]. Le et al. [76] fabricated graphene electrodes with inkjet printing of graphene oxide, followed by thermal reduction and found that the electrochemical performance of inkjet printing is favorably comparable to other methods. Graphene oxide dispersed in water was used as an ink to develop a graphene-based inkjet printed supercapacitor [77]. The specific capacitance of up to 192 F/g and the loss of capacitance less than 5% was observed after the repeated bending cycles of the device. SWNT inks were used through an inkjet printer on a cloth fabric to produce thin-film electrodes. These films were then sandwiched between polymer electrolytes to develop supercapacitors. The performance of the printed SWNT supercapacitor was remarkably improved in terms of its specific capacitance of 138 F/g, power density and energy density, with the addition of RuO_2 nanowires [78].

3.5 3D printing

The 3D printing technology could produce low-cost 3D printed platforms for various applications. As electrochemical 3D systems have recently been explored, there has been a particular focus on the development of supercapacitors [79, 80]. The 3D technique is one effective way to improve the overall energy performance of stretchable supercapacitors without compromising their mechanical properties [81], and is also famous to improve the energetic areal performance of micro-supercapacitors [82]. Zhu et al. [83] reported the fabrication of 3D printed aerogel for supercapacitor applications using the 3D printing fabrication method. The developed supercapacitor exhibited exceptional capacitive retention and power densities. A highly flexible electrochemical double-layer capacitor was developed in a single continuous manufacturing process using the 3D printing method [84]. All the components of the supercapacitor were fabricated in a grid pattern. Electrochemical performance and flexibility of the 3D printed supercapacitors were investigated using the mechanical bending tests, which were found excellent with the retention of 54-58% of its initial capacitance at 50 mV s-1 scan rate. Moreover, they proved that the 3D printing technique has good reproducibility and can develop various electronic devices.

3.6 Spray coating

The spray coating technique is generally preferred for large scale production because there is no restraint in the size of the substrate and polymer utilization. This technique is a substitute for the conventional spin coating method [85]. An aerosol is formed, as the printing ink comes out through a nozzle [86]. However, spray coating application for active materials is faced with issues like high film thickness and roughness [87]. Thus, researchers are concerned about improving the morphology of an active layer by means of solvents with high boiling points [88].

Li et al. [89] used the spray coating technique to deposit the silver electrode. They altered the morphology of spray-coated silver electrodes by hydrochloric acid Solvent Vapor Annealing (SVA). They provided a promising technique to prepare large-scale PSC for the fabrication of printed electronics. In another study, Sprayable ink based on activated carbon and single-layer graphene flakes was reported in [90]. Ink deposition through spray coating enhanced the electrolyte accessibility to the electrode surface area. The superior rate capability with the specific energies of 31.5 Wh/kg and 12.5 Wh/kg was displayed at specific powers of 150 W/kg and 30 kW/kg, respectively.

There are many other techniques deployed by scholars to fabricate supercapacitor electrodes that include screen printing [91, 92], vacuum filtration [93, 94], electroless deposition [95], electrospinning [96], blade coating [97], carbonization [98], sol–gel [99, 100] etc.

4. Performance evaluation

4.1 Governing equations

Chen & Dai [101] explained the electrochemical supercapacitors governing equations for performance evaluation when charge cumulates at electrode and electrolyte interface. The capacitance (C) being represented by Eq. (1):

$$C = \dot{\in}_r \dot{\in}_0 \left(\frac{A}{D}\right) \tag{1}$$

Where $\acute{\varepsilon}_r$ is the electrolyte dielectric constant, $\acute{\varepsilon}_0$ is the dielectric constant of the vacuum, d is the effective thickness of the double layer (charge separation distance), and A is the electrode surface area. The overall capacitance (C_c) of the entire device or cell and the normalized capacitance (C_{NOR}) of an electrode or electrode materials can be expressed via Eqs. (2) and (3):

$$\frac{1}{C_c} = \frac{1}{C_{(p+)}} + \frac{1}{C_{(n-)}} \tag{2}$$

$$C_{NOR} = \frac{C_i}{P_i} \tag{3}$$

where $C_{(p+)}$ and $C_{(n-)}$ represent the capacitance of the positive electrode and negative electrodes' capacitance, respectively. C_i is the capacitance of the device or electrode materials (individual electrode). P_i could be the parameter either related with the weight (resulting in gravimetric capacitance, F/g), the area (resulting in aerial capacitance, F/cm^2) or the volume (resulting in volumetric capacitance, F/cm^3) of the electrode or electrode materials to achieve the normalized capacitance values. In the case of a symmetric device, the capacitance of the positive electrode $C_{(p+)}$ must equal that of the negative one $C_{(n-)}$. Thus, the capacitance of the complete cell or device is half of the capacitance of each individual electrode (C_i), that is, $C_i = C_{(p+)} = C_{(n-)}$

$$C_c = \frac{C_i}{2} \tag{4}$$

The energy density (E) and power density (P) can be calculated using Eqs. (5) and (6):

$$E = 0.5\ CV^2 \tag{5}$$

$$P = \frac{V^2}{4R_{ESR}} = \frac{E}{t_{discharge}} \tag{6}$$

Where V is the operating voltage, R_{ESR} is the equivalent series resistance of the device, and t $_{discharge}$ is the discharge time.

4.2 Cell setup for supercapacitor testing

The supercapacitor testing is carried out through an electrochemical workstation (EW), which contains electronic hardware units of potentiostat and galvanostat. It also contains a frequency response analyzing unit as an option to characterize electrochemical impedance spectroscopy. The user-specified counter electrode (C_E) potential is precisely controlled against the working electrode (W_E) potential, and the current response being observed in potentiostatic mode. In contrast, the flow of current is accurately maintained between working and counter electrodes in galvanostatic mode.

The cell setup for supercapacitor performance output characterization consists of three and two-electrode configurations. Two electrode test cell design gives the real output performance of the supercapacitor device, i.e. the obtained results include both electrode contributions towards the capacitance value. On the other hand, the latter is used to derive effective electrode material performance results. This means the results that are acquired through a three-electrode setup are exclusively due to the working electrode (the electrode understudy) and do not include any invasion from another electrode.

It is easy to extract output information for individual electrodes while using two-electrode cell assembly for symmetrical devices, but difficult to obtain such information accurately for the asymmetric device since using two-electrode methods is not feasible. The main reason for not recommending a two-electrode configuration for asymmetric devices is that accurate contribution to overall capacitance values arising from two different electrodes cannot be distinguished. A three-electrode cell should be used in such scenarios if individual electrode performance is studied in an asymmetric supercapacitor [102]. The three-cell setup consists of a working electrode, a counter electrode, and a reference electrode, while in the two-electrode setup, there are two electrodes; one is positive, and the other is negative, separated by electrolyte-soaked porous membrane.

4.3 Electrochemical techniques employed for supercapacitor characterization

The critical performance parameters of a supercapacitor are measured using three main electrochemical techniques, which include cyclic voltammetry (CV), Galvanostatic Charge/Discharge (GCD, also known as constant current charge/discharge) and Electrochemical Impedance Spectroscopy (EIS). Electrochemical workstation is used for all these techniques to quantify key factors such as voltage, current, time, equivalent series resistance and capacitance in three or two-electrode configurations. From the measurement results, the power and energy values of the tested device can be calculated through mathematical Equations [103].

4.3.1 Cyclic voltammetry (CV)

Cyclic voltammetry is a versatile, dynamic electrochemical method for evaluating the electrochemical capabilities of a device or material through electrode kinetics and charge storage mechanism happening at the electrode/electrolyte interface. Both three and two electrode configurations can be used in a CV experiment to measure various parameters. Working electrode potential is measured against a reference electrode which maintains a constant potential. Distinct shape CV graphs are generated (rectangular/quasi-rectangular) as shown in **Figure 1a**, with time-dependent current on the vertical axis and predetermined voltage window on the horizontal axis, when a fixed rate linear varying voltage is swept between two predetermined upper and lower electric potential values. The scan is forwarded and reversed

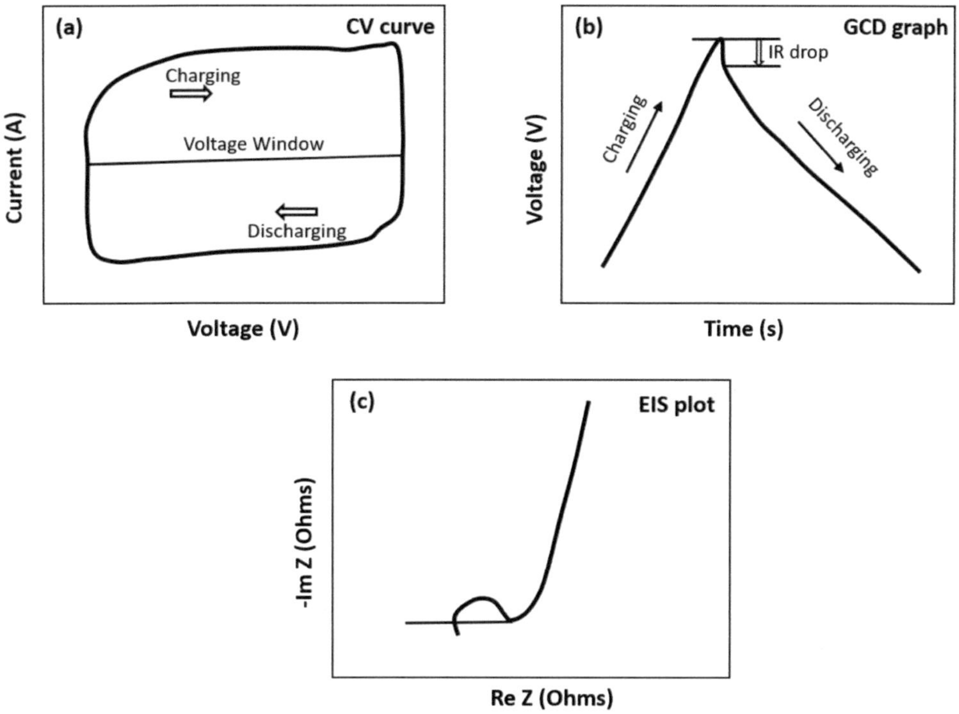

Figure 1.
Typical profiles of a) CV, b) GCD, and c) EIS of a supercapacitor.

between this voltage bracket, also known as operating potential or voltage window. This voltage window value depends upon the type of electrolyte used, as the voltage range should not surpass the stable voltage operating window of the electrolyte.

Exceeding the voltage limit will start the electrolysis process of an electrolyte, and this reaction will govern the cell chemistry, thus resulting in deteriorated cell performance [104]. Different scan rates (mV/s) are also employed during CV analysis; this voltage/potential change speeds during the experimental setup. The scan rate has a considerable impact on capacitance values and graphs obtained. It is mainly related to electrode kinetics, as the decrease in specific capacitance is associated with limited ion transfer due to increased scan rate [102]. At slower scan rates, the CV graphs also show better rectangular charts. Eq. (7) can be used to calculate the capacitance values of a super cache capacitor when the CV technique is used in two-electrode assemblies [105].

$$C_{wt} = \frac{4\int_{V_1}^{V_n} i\, dV}{ms\Delta V} \qquad (7)$$

where is the integrated area of the CV curve, V is (2 × the voltage window, from E1 to E2 = (|E2 -E1|) in volts (V), s is the scan rate (V/s), m represents the mass (g) of active materials on both electrodes [105].

The multiplication factor of 4 should be replaced by 2 in the three-cell setup to calculate the gravimetric capacitance.

4.3.2 Galvanostatic charge–discharge (GCD)

GCD is another heavily deployed technique to calculate capacitance, power, energy densities, the equivalent series resistance of the supercapacitor device or material. It also helps in identifying parameters such as the life cycle and the stability of the device during that period [103]. In this method, a constant positive current is applied to the working electrode at the start, so it is charged to its peak voltage value that is specified in the voltage window, and voltage response is recorded against charging time. Once fully charged, a constant negative current is applied to the device, so it is discharged to its lower voltage value, and voltage response against discharge time are noted [105]. The GCD profile formed for a supercapacitor is triangular, as shown in **Figure 1b**. Eq. (8) can be used to calculate areal capacitance (F/cm^2) in a two-electrode assembly [105];

$$C_A = \frac{4I}{A\left(\frac{\Delta E}{\Delta t}\right)} \tag{8}$$

where is the current in amperes (A), represents the slope of the discharge curve from IR drop, and A is the area in cm^2 of the two electrodes.

4.3.3 Electrochemical impedance spectroscopy (EIS)

This technique provides beneficial information such as impedance, frequency response on capacitance, electrode/electrolyte interface related phenomena such as charge transfer, and mass transport. EIS is also known as AC impedance spectroscopy and dielectric spectroscopy. In this method, a low amplitude ACs, signal (voltage-potentiostatic or current control-galvanostatic) is superimposed on a steady-state signal over a variable frequency range, and its effect is studied on the impedance of the device [102]. The data is usually presented on a Nyquist plot (**Figure 1c**) where the real and imaginary impedance of the device is shown. Calculating capacitance via EIS test using traditional method basically relies on imaginary complex impedance part Im(Z) and is shown in Eq. (9) [103].

$$C_{Tf} = -\frac{1}{2\pi f\, I\, m\, (Z)} \tag{9}$$

Where f is usually the lowest applied frequency value.

5. Conclusion

Supercapacitors provide promising solution to the future energy storage devices. Various materials and fabrication options are available for the development of supercapacitors. Key performance parameters and their assessment criteria have been reviewed in this chapter.

Extensive research has been conducted in the past decade on supercapacitors with quite significant progress. Although inspiring results have been achieved in this field; yet the energy storage devices still require further enhancement in device fabrication techniques to achieve uniformity, scalability, and consistency in performance evaluation to satisfy commercial demands. It is expected that future research and developments will address these challenges.

Acknowledgements

The authors would like to thank the Australian Research Council (ARC) for their support for this research project through grant LP110100455.

Conflict of interest

The authors declare no conflict of interest.

Author details

Imtiaz Ahmed Khan[1*], Litty Thekkekara[1], Sania Waqar[2], Nauman Choudhry[2] and Sabu John[1]

1 School of Engineering, RMIT University, Australia

2 School of Fashion and Textiles, RMIT University, Australia

*Address all correspondence to: imtikhan20@hotmail.com

References

[1] Wang Y et al. Recent progress in carbon-based materials for supercapacitor electrodes: A review. Journal of Materials Science. 2021;**56**(1):173-200

[2] Yu A, Chen Z, Maric R, Zhang L, Zhang J, Yan J. Electrochemical supercapacitors for energy storage and delivery: Advanced materials, technologies and applications. Applied Energy. 2015;**153**:1-2

[3] Forouzandeh P, Kumaravel V, Pillai SC. Electrode materials for supercapacitors: A review of recent advances. Catalysts. 2020;**10**(9):969

[4] Chen GZ. Supercapacitor and supercapattery as emerging electrochemical energy stores. International Materials Reviews. 2017; **62**(4):173-202

[5] Conway BE. Electrochemical Supercapacitors: Scientific Fundamentals and Technological Applications. New York: Springer Science & Business Media; 2013

[6] Kötz R, Carlen M. Principles and applications of electrochemical capacitors. Electrochimica Acta. 2000; **45**(15-16):2483-2498

[7] Guan M, Liao W. Characteristics of energy storage devices in piezoelectric energy harvesting systems. Journal of Intelligent Material Systems and Structures. 2008;**19**(6):671-680

[8] Pandolfo AG, Hollenkamp AF. Carbon properties and their role in supercapacitors. Journal of Power Sources. 2006;**157**(1):11-27

[9] Burke A. Ultracapacitors: Why, how, and where is the technology. Journal of Power Sources. 2000;**91**(1): 37-50

[10] Zhang Y et al. Progress of electrochemical capacitor electrode materials: A review. International Journal of Hydrogen Energy. 2009; **34**(11):4889-4899

[11] Zhang X, Wang X, Jiang L, Wu H, Wu C, Su J. Effect of aqueous electrolytes on the electrochemical behaviors of supercapacitors based on hierarchically porous carbons. Journal of Power Sources. 2012;**216**:290-296

[12] Wang H et al. Interconnected carbon nanosheets derived from hemp for ultrafast supercapacitors with high energy. ACS Nano. 2013;**7**(6): 5131-5141

[13] Abdulhakeem B et al. Morphological characterization and impedance spectroscopy study of porous 3D carbons based on graphene foam-PVA/phenol-formaldehyde resin composite as an electrode material for supercapacitors. RSC Advances. 2014; **4**(73):39066-39072

[14] An KH et al. Supercapacitors using single-walled carbon nanotube electrodes. Advanced Materials. 2001; **13**(7):497-500

[15] Cottineau T, Toupin M, Delahaye T, Brousse T, Bélanger D. Nanostructured transition metal oxides for aqueous hybrid electrochemical supercapacitors. Applied Physics A. 2006;**82**(4): 599-606

[16] Liu Y, Zhang B, Xiao S, Liu L, Wen Z, Wu Y. A nanocomposite of MoO3 coated with PPy as an anode material for aqueous sodium rechargeable batteries with excellent electrochemical performance. Electrochimica Acta. 2014;**116**:512-517

[17] Beidaghi M, Wang C. Micro-supercapacitors based on three dimensional interdigital polypyrrole/C-MEMS electrodes. Electrochimica Acta. 2011;**56**(25):9508-9514

[18] Choudhry NA, Arnold L, Rasheed A, Khan IA, Wang L. Textronics—A review of textile-based wearable electronics. Advanced Engineering Materials. 2021; **23**(11):2100469

[19] Yun YS, Park HH, Jin H-J. Pseudocapacitive effects of N-doped carbon nanotube electrodes in supercapacitors. Materials. 2012;**5**(7): 1258-1266

[20] Raza W et al. Recent advancements in supercapacitor technology. Nano Energy. 2018;**52**:441-473

[21] Wang F et al. Latest advances in supercapacitors: From new electrode materials to novel device designs. Chemical Society Reviews. 2017;**46**(22): 6816-6854

[22] Bansal RC, Goyal M. Activated Carbon Adsorption. Boca Raton: CRC Press; 2005

[23] Yasmin A, Luo J-J, Daniel IM. Processing of expanded graphite reinforced polymer nanocomposites. Composites Science and Technology. 2006;**66**(9):1182-1189

[24] Duplock EJ, Scheffler M, Lindan PJ. Hallmark of perfect graphene. Physical Review Letters. 2004;**92**(22):225502

[25] Novoselov KS et al. Two-dimensional gas of massless Dirac fermions in graphene. Nature. 2005; **438**(7065):197-200

[26] Zhang L et al. Porous 3D graphene-based bulk materials with exceptional high surface area and excellent conductivity for supercapacitors. Scientific Reports. 2013;**3**(1):1-9

[27] Bonaccorso F, Lombardo A, Hasan T, Sun Z, Colombo L, Ferrari AC. Production and processing of graphene and 2d crystals. Materials Today. 2012; **15**(12):564-589

[28] Chiu PL et al. Microwave-and nitronium ion-enabled rapid and direct production of highly conductive low-oxygen graphene. Journal of the American Chemical Society. 2012; **134**(13):5850-5856

[29] Wei D, Liu Y, Wang Y, Zhang H, Huang L, Yu G. Synthesis of N-doped graphene by chemical vapor deposition and its electrical properties. Nano Letters. 2009;**9**(5):1752-1758

[30] Basnayaka PA, Ram MK, Stefanakos L and Kumar A. Graphene/polypyrrole nanocomposite as electrochemical supercapacitor electrode: Electrochemical impedance studies. 2013;**2**(2):30869

[31] Li X, Wang Z, Qiu Y, Pan Q, Hu P. 3D graphene/ZnO nanorods composite networks as supercapacitor electrodes. Journal of Alloys and Compounds. 2015; **620**:31-37

[32] Zhang LL, Zhao X. Carbon-based materials as supercapacitor electrodes. Chemical Society Reviews. 2009;**38**(9): 2520-2531

[33] El-Kady MF, Kaner RB. Scalable fabrication of high-power graphene micro-supercapacitors for flexible and on-chip energy storage. Nature Communications. 2013;**4**(1):1-9

[34] Huang L, Wu B, Yu G, Liu Y. Graphene: learning from carbon nanotubes. Journal of Materials Chemistry. 2011;**21**(4):919-929

[35] Dresselhaus G, Dresselhaus MS, Saito R. Physical Properties of Carbon Nanotubes. London: Imperial College Press. 1998

[36] Hutchison J et al. Double-walled carbon nanotubes fabricated by a hydrogen arc discharge method. Carbon. 2001;**39**(5):761-770

[37] Scott CD, Arepalli S, Nikolaev P, Smalley RE. Growth mechanisms for single-wall carbon nanotubes in a laser-

ablation process. Applied Physics A. 2001;**72**(5):573-580

[38] Bronikowski MJ, Willis PA, Colbert DT, Smith K, Smalley RE. Gas-phase production of carbon single-walled nanotubes from carbon monoxide via the HiPco process: A parametric study. Journal of Vacuum Science & Technology A: Vacuum, Surfaces, and Films. 2001;**19**(4): 1800-1805

[39] Guellati O et al. Influence of ethanol in the presence of H2 on the catalytic growth of vertically aligned carbon nanotubes. Applied Catalysis A: General. 2012;**423**:7-14

[40] Pandey A, Chouhan RS, Gurbuz Y, Niazi JH, Qureshi A. S. cerevisiae whole-cell based capacitive biochip for the detection of toxicity of different forms of carbon nanotubes. Sensors and Actuators B: Chemical. 2015;**218**: 253-260

[41] Bello A et al. High-performance symmetric electrochemical capacitor based on graphene foam and nanostructured manganese oxide. AIP Advances. 2013;**3**(8):082118

[42] Nagarajan N, Humadi H, Zhitomirsky I. Cathodic electrodeposition of MnOx films for electrochemical supercapacitors. Electrochimica Acta. 2006;**51**(15): 3039-3045

[43] Toupin M, Brousse T, Bélanger D. Charge storage mechanism of MnO2 electrode used in aqueous electrochemical capacitor. Chemistry of Materials. 2004;**16**(16):3184-3190

[44] Bello A, Makgopa K, Fabiane M, Dodoo-Ahrin D, Ozoemena KI, Manyala N. Chemical adsorption of NiO nanostructures on nickel foam-graphene for supercapacitor applications. Journal of Materials Science. 2013;**48**(19):6707-6712

[45] Wu M-S, Hsieh H-H. Nickel oxide/hydroxide nanoplatelets synthesized by chemical precipitation for electrochemical capacitors. Electrochimica Acta. 2008;**53**(8): 3427-3435

[46] Srinivasan V, Weidner JW. An electrochemical route for making porous nickel oxide electrochemical capacitors. Journal of the Electrochemical Society. 1997;**144**(8): L210

[47] Hyun T-S, Kang J-E, Kim H-G, Hong J-M, Kim I-D. Electrochemical properties of MnO x–RuO2 nanofiber mats synthesized by Co-electrospinning. Electrochemical and Solid State Letters. 2009;**12**(12):A225

[48] Ahn YR, Song MY, Jo SM, Park CR, Kim DY. Electrochemical capacitors based on electrodeposited ruthenium oxide on nanofibre substrates. Nanotechnology. 2006;**17**(12):2865

[49] Liu T, Pell WG, Conway BE. Self-discharge and potential recovery phenomena at thermally and electrochemically prepared RuO2 supercapacitor electrodes. Electrochimica Acta. 1997;**42**(23-24): 3541-3552

[50] Chang J-K, Wu C-M, Sun I-W. Nano-architectured Co (OH) 2 electrodes constructed using an easily-manipulated electrochemical protocol for high-performance energy storage applications. Journal of Materials Chemistry. 2010;**20**(18):3729-3735

[51] Mendoza-Sánchez B, Brousse T, Ramirez-Castro C, Nicolosi V, Grant PS. An investigation of nanostructured thin film α-MoO3 based supercapacitor electrodes in an aqueous electrolyte. Electrochimica Acta. 2013;**91**:253-260

[52] Aravinda L, Bhat U, Bhat BR. Binder free MoO3/multiwalled carbon nanotube thin film electrode for high

energy density supercapacitors. Electrochimica Acta. 2013;**112**:663-669

[53] Zhou Y et al. Advanced asymmetric supercapacitor based on conducting polymer and aligned carbon nanotubes with controlled nanomorphology. Nano Energy. 2014;**9**:176-185

[54] Snook GA, Kao P, Best AS. Conducting-polymer-based supercapacitor devices and electrodes. Journal of Power Sources. 2011;**196**(1): 1-12

[55] Roberts ME, Wheeler DR, McKenzie BB, Bunker BC. High specific capacitance conducting polymer supercapacitor electrodes based on poly (tris (thiophenylphenyl) amine). Journal of Materials Chemistry. 2009; **19**(38):6977-6979

[56] Ma Z, Zheng R, Liu Y, Ying Y, Shi W. Carbon nanotubes interpenetrating MOFs-derived Co-Ni-S composite spheres with interconnected architecture for high performance hybrid supercapacitor. Journal of Colloid and Interface Science. 2021;**602**: 627-635

[57] Jurewicz K, Delpeux S, Bertagna V, Beguin F, Frackowiak E. Supercapacitors from nanotubes/ polypyrrole composites. Chemical Physics Letters. 2001;**347**(1-3):36-40

[58] Frackowiak E, Khomenko V, Jurewicz K, Lota K, Béguin F. Supercapacitors based on conducting polymers/nanotubes composites. Journal of Power Sources. 2006;**153**(2): 413-418

[59] Wang G, Zhang L, Zhang J. A review of electrode materials for electrochemical supercapacitors. Chemical Society Reviews. 2012;**41**(2): 797-828

[60] Kim BK, Sy S, Yu A, Zhang J. Electrochemical supercapacitors for energy storage and conversion. In: Handbook of Clean Energy Systems. USA: John Wiley & Sons, Ltd. 2015. pp. 1-25

[61] Winter M, Brodd RJ. What are batteries, fuel cells, and supercapacitors? Chemical Reviews. 2004;**104**(10):4245-4270

[62] Verma KD, Sinha P, Banerjee S, Kar KK, Ghorai MK. Characteristics of separator materials for supercapacitors. In: Handbook of Nanocomposite Supercapacitor Materials I. Switzerland AG: Springer; 2020. pp. 315-326

[63] Zhang K et al. Facile synthesis of high density carbon nanotube array by a deposition-growth-densification process. Carbon. 2017;**114**:435-440

[64] Yao Y et al. Temperature-mediated growth of single-walled carbon-nanotube intramolecular junctions. Nature Materials. 2007;**6**(4):283-286

[65] Das R, Shahnavaz Z, Ali ME, Islam MM, Abd Hamid SB. Can we optimize arc discharge and laser ablation for well-controlled carbon nanotube synthesis? Nanoscale Research Letters. 2016;**11**(1):1-23

[66] Prasek J et al. Methods for carbon nanotubes synthesis. Journal of Materials Chemistry. 2011;**21**(40): 15872-15884

[67] Li S et al. Enhancement of carbon nanotube fibres using different solvents and polymers. Composites Science and Technology. 2012;**72**(12):1402-1407

[68] Guo W, Liu C, Sun X, Yang Z, Kia HG, Peng H. Aligned carbon nanotube/polymer composite fibers with improved mechanical strength and electrical conductivity. Journal of Materials Chemistry. 2012;**22**(3): 903-908

[69] Peng H et al. Electrochromatic carbon nanotube/polydiacetylene

nanocomposite fibres. Nature Nanotechnology. 2009;**4**(11):738-741

[70] Mahltig B, Fiedler D, Fischer A, Simon P. Antimicrobial coatings on textiles–modification of sol–gel layers with organic and inorganic biocides. Journal of Sol-Gel Science and Technology. 2010;**55**(3): 269-277

[71] Hu L et al. Stretchable, porous, and conductive energy textiles. Nano Letters. 2010;**10**(2):708-714

[72] Su F, Miao M. Asymmetric carbon nanotube–MnO2 two-ply yarn supercapacitors for wearable electronics. Nanotechnology. 2014; **25**(13):135401

[73] Paleo A, Staiti P, Brigandì A, Ferreira F, Rocha A, Lufrano F. Supercapacitors based on AC/MnO2 deposited onto dip-coated carbon nanofiber cotton fabric electrodes. Energy Storage Materials. 2018;**12**: 204-215

[74] Ju H, Lee JK, Lee J, Lee J. Fast and selective Cu2O nanorod growth into anodic alumina templates via electrodeposition. Current Applied Physics. 2012;**12**(1):60-64

[75] Sajedi-Moghaddam A, Rahmanian E, Naseri N. Inkjet-printing technology for supercapacitor application: Current state and perspectives. ACS Applied Materials & Interfaces. 2020;**12**(31):34487-34504

[76] Le LT, Ervin MH, Qiu H, Fuchs BE, Lee WY. Graphene supercapacitor electrodes fabricated by inkjet printing and thermal reduction of graphene oxide. Electrochemistry Communications. 2011;**13**(4):355-358

[77] Ervin MH, Le LT, Lee WY. Inkjet-printed flexible graphene-based supercapacitor. Electrochimica Acta. 2014;**147**:610-616

[78] Chen P, Chen H, Qiu J, Zhou C. Inkjet printing of single-walled carbon nanotube/RuO 2 nanowire supercapacitors on cloth fabrics and flexible substrates. Nano Research. 2010;**3**(8):594-603

[79] Zhao C, Wang C, Gorkin Iii R, Beirne S, Shu K, Wallace GG. Three dimensional (3D) printed electrodes for interdigitated supercapacitors. Electrochemistry Communications. 2014;**41**:20-23

[80] Foster CW et al. 3D printed graphene based energy storage devices. Scientific Reports. 2017;7(1):1-11

[81] Chang P, Mei H, Tan Y, Zhao Y, Huang W, Cheng L. A 3D-printed stretchable structural supercapacitor with active stretchability/flexibility and remarkable volumetric capacitance. Journal of Materials Chemistry A. 2020; **8**(27):13646-13658

[82] Li X, Li H, Fan X, Shi X, Liang J. 3D-printed stretchable micro-supercapacitor with remarkable areal performance. Advanced Energy Materials. 2020;**10**(14):1903794

[83] Zhu C et al. Supercapacitors based on three-dimensional hierarchical graphene aerogels with periodic macropores. Nano Letters. 2016;**16**(6): 3448-3456

[84] Areir M, Xu Y, Harrison D, Fyson J. 3D printing of highly flexible supercapacitor designed for wearable energy storage. Materials Science and Engineering: B. 2017;**226**:29-38

[85] Kang J-W et al. Fully spray-coated inverted organic solar cells. Solar Energy Materials and Solar Cells. 2012;**103**: 76-79

[86] Krebs FC. Fabrication and processing of polymer solar cells: A review of printing and coating

techniques. Solar Energy Materials and Solar Cells. 2009;**93**(4):394-412

[87] Girotto C, Rand BP, Genoe J, Heremans P. Exploring spray coating as a deposition technique for the fabrication of solution-processed solar cells. Solar Energy Materials and Solar Cells. 2009;**93**(4):454-458

[88] Green R, Morfa A, Ferguson A, Kopidakis N, Rumbles G, Shaheen S. Performance of bulk heterojunction photovoltaic devices prepared by airbrush spray deposition. Applied Physics Letters. 2008;**92**(3):17

[89] Li S, Zheng Y, Cheng J, Tu M, Yu J. Effect of hydrochloric acid solvent vapor annealing on spray coated silver electrode. Journal of Materials Science: Materials in Electronics. 2014;**25**(11): 5013-5019

[90] Garakani MA et al. Scalable spray-coated graphene-based electrodes for high-power electrochemical double-layer capacitors operating over a wide range of temperature. Energy Storage Materials. 2021;**34**:1-11

[91] Xiong C, Li M, Han Q, Zhao W, Dai L, Ni Y. Screen printing fabricating patterned and customized full paper-based energy storage devices with excellent photothermal, self-healing, high energy density and good electromagnetic shielding performances. Journal of Materials Science & Technology. 2022;**97**:190-200

[92] Alam A, Saeed G, Lim S. Screen-printed activated carbon/silver nanocomposite electrode material for a high performance supercapacitor. Materials Letters. 2020;**273**:127933

[93] Tan H, Xiao D, Navik R, Goto M, Zhao Y. Fabrication of graphene/polyaniline nanofiber multilayer composite for supercapacitor electrodes via layer-by-layer vacuum filtration. Journal of Materials Science: Materials in Electronics. 2020;**31**(21):18569-18580

[94] Huang Y-L, Bian S-W. Vacuum-filtration assisted layer-by-layer strategy to design MXene/carbon nanotube@ MnO 2 all-in-one supercapacitors. Journal of Materials Chemistry A. 2021; **9**(37):21347-21356

[95] Tantawy N, Heakal FE-T, Ahmed S. Synthesis of worm-like binary metallic active material by electroless deposition approach for high-performance supercapacitor. Journal of Energy Storage. 2020;**31**:101625

[96] Bhattacharya S et al. High-conductivity and high-capacitance electrospun fibers for supercapacitor applications. ACS Applied Materials & Interfaces. 2020;**12**(17):19369-19376

[97] Abdolhosseinzadeh S, Heier J, Zhang C. Coating porous MXene films with tunable porosity for high-performance solid-state supercapacitors. ChemElectroChem. 2021;**8**(10):1911-1917

[98] Xu Z-X et al. Benign-by-design N-doped carbonaceous materials obtained from the hydrothermal carbonization of sewage sludge for supercapacitor applications. Green Chemistry. 2020; **22**(12):3885-3895

[99] Kumar A. Sol gel synthesis of zinc oxide nanoparticles and their application as nano-composite electrode material for supercapacitor. Journal of Molecular Structure. 2020;**1220**:128654

[100] Uke SJ, Mardikar SP, Bambole DR, Kumar Y, Chaudhari GN. Sol-gel citrate synthesized Zn doped MgFe2O4 nanocrystals: A promising supercapacitor electrode material. Materials Science for Energy Technologies. 2020;**3**:446-455

[101] Chen T, Dai L. Flexible supercapacitors based on carbon

nanomaterials. Journal of Materials Chemistry A. 2014;**2**(28):10756-10775

[102] Yu A, Chabot V, Zhang J. Electrochemical Supercapacitors for Energy Storage and Delivery: Fundamentals and Applications. New York: Taylor & Francis; 2013

[103] Zhang S, Pan N. Supercapacitors performance evaluation. Advanced Energy Materials. 2015;**5**(6):1401401

[104] Smith L, Dunn B. Opening the window for aqueous electrolytes. Science. 2015;**350**(6263):918-918

[105] Moussa M, El-Kady MF, Zhao Z, Majewski P, Ma J. Recent progress and performance evaluation for polyaniline/graphene nanocomposites as supercapacitor electrodes. Nanotechnology. 2016;**27**(44):442001

Section 2

Biomass Based Materials

Chapter 2

Biomass-Based Carbon Electrodes in the Design of Supercapacitors: An Electrochemical Point of View

Antony Bazan-Aguilar, Elvis O. López, Miguel Ponce-Vargas and Angélica M. Baena-Moncada

Abstract

The urgent demand of sustainable long-lasting batteries has fostered the improvement of extended-use technologies e.g., Li-ion batteries, as well as the development of alternative energy storage strategies like supercapacitors. In this context, new carbon-based materials were developed to attain higher electrochemical performances, even though several of these materials are not obtained by eco-friendly methods and/or in a considerable amount for practical purposes. However, up-to-date reports stand out the scopes achieved by biomass-based carbon materials as energy storage electrodes combining outstanding physicochemical and electrochemical properties with low-pollutant and low-cost production. On this basis, this chapter will expose several aspects of the synthesis of carbon-based electrodes from biomass, focusing on the influence of their surface properties: porosity, crystallinity, and morphology on their electrochemical performance in supercapacitors.

Keywords: Biomass-based carbon electrodes, Energy storage device, Surface properties, Electrochemical performance, Supercapacitors

1. Introduction

Nowadays, the search for new eco-friendly energy systems is a priority to mitigate the global impact associated with fossil fuel energy consumption [1–4]. At the same time, novel electronic devices need to be developed to produce more efficient energy storage systems with higher capacity and longer average lifetimes [5]. In this regard, electrical double layer capacitors (EDLC), pseudocapacitors [6] and flexible solid-state supercapacitors (FSSC) [7, 8], are able to cover the above-mentioned demands. Particularly, carbon-based capacitors exhibit significant advantages, such as high-power density, low weight and flexibility, in contrast to conventional graphite-based systems [9]. To achieve a better capacitance, electrodes based in porous carbon materials can be employed, given their interesting morphological features, like high specific surface area (S_{BET}), defined-porosity, and hierarchical arrangement [10]. Recently, the use of biomass as precursor material has allowed the design of carbon-based energy storage systems with outstanding electrochemical and mechanical properties [11, 12].

The energy storage mechanism in these materials consists in the accumulation of electrostatic charges on their surface (**Figure 1**), [13] implying that, density of stored charges depends on the material morphology (roughness and pores), and the size of the

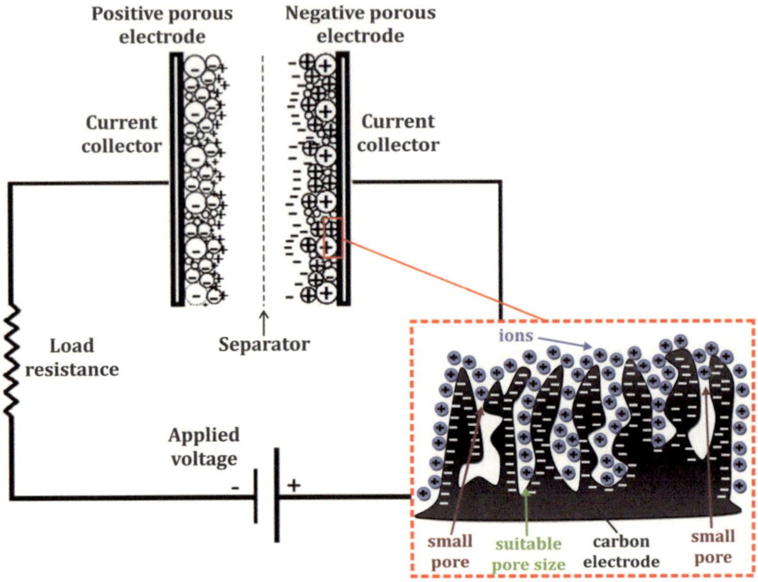

Figure 1.
Schematic representation of a loaded EDLC supercapacitor and pore accessibility (inset) with respect to the electrolyte. Figure adapted from [13, 14].

involved electrolyte ions [14]. **Figure 1** shows the interplay between electrode porosity and ion-electrolyte size for the electrostatic charge accumulation, in terms of pore accessibility. In this way, activated carbons mainly exhibit three pore sizes [15]: micro (< 2 nm), meso (2–50 nm) and macro (> 50 nm), being the micropores and mesopores the ones that most contribute to the increase of the capacitance [16]. Nevertheless, even though carbon-based materials have high surface areas (1000–2000 $m^2 g^{-1}$), they still have low specific capacitances due to their limited mesoporosity [14].

The carbons from biomass can be obtained from various natural sources, such as: garlic skin, bamboo, rice husk, eucalyptus-bark, lignin, cellulose, orange peel, etc. [6, 17–22], making biomass a sustainable source, also available in large quantities from industrial waste [18]. Furthermore, the carbons obtained from biomass present a diversified class of fibrous and porous structures [22], with capacitances ranging from 100 to 430 F g^{-1} [16, 23]. However, these values vary according to the carbon activation method, which is strongly related to the resulting electrode pore-size. Such processes of carbon activation from biomass can be divided into two categories: i) physical activation and ii) chemical activation.

1.1 Physical activation processes

In this process the biomass undergoes a pyrolysis treatment at temperatures between 600 and 900°C in inert atmosphere. Then, the material is commonly exposed to an oxidizing atmosphere of CO_2, carried out at temperatures between 600 and 1200°C. Yu et al. used this method to obtain activated carbon from cattail biomass [24], resulting in an activated carbon with a surface area of 441 $m^2 g^{-1}$ and a specific capacitance of 126 F g^{-1} (current density of 0.5 A g^{-1} in KOH 6 mol L^{-1} electrolyte) [18].

1.2 Chemical activation processes

In this process, the carbon precursor (biomass) is treated (soaked) with chemical activators, such as: KOH, NaOH, H_3PO_4, $ZnCl_2$, H_2SO_4, among others.

Subsequently, the biomass is carbonized at temperatures between 400 and 900°C [25]. During this activation, redox processes and substitution of large particles take place, leading to the desired high porosity. In addition, physical activation occurs due to the interaction of the reaction products such as: H_2, H_2O, CO and CO_2. Remarkably, these molecules also contribute to the pore formation.

Using KOH as a chemical activator, K^0 (metallic) is produced which occupies interstitial positions in the carbon structure, inducing expansion in the material as well as producing a high microporosity. A proposed mechanism of the activation process using KOH is shown in the following redox reactions [26]:

$$6KOH + 2C \rightarrow 2K + 3H_2 + 2K_2CO_3 \tag{1}$$

$$2K_2CO_3 \rightarrow 2K_2O + 2CO_2 \tag{2}$$

$$K_2CO_3 + 2C \rightarrow 2K + 3CO \tag{3}$$

$$CO_2 + C \rightarrow 2CO \tag{4}$$

$$C + K_2O \rightarrow 2K + CO \tag{5}$$

Diverse works have used the chemical activation process from KOH, Barzegar et al. was one of them and produced mesoporous carbon from coconut shell, obtaining a specific surface area of 1416 $m^2\ g^{-1}$ and a specific capacity of 186 $F\ g^{-1}$ [27]. Zhang et al. used bamboo to obtain mesoporous carbon from KOH as an activator. Zhang obtained a high specific surface area of 2221 $m^2\ g^{-1}$ and a specific capacitance of 293 $F\ g^{-1}$ at 0.5 $A\ g^{-1}$ in KOH 3.0 mol L^{-1} [18]. Yin et al. produced activated carbon from coconut fibers with KOH as a chemical activator. This product exhibits a surface area of 2898 $m^2\ g^{-1}$ (pore volume of 1.59 $cm^3\ g^{-1}$, i.e. 30% of mesoporosity) and specific capacitance of 266 $F\ g^{-1}$ at 0.1 $A\ g^{-1}$ in KOH 6.0 mol L^{-1} [28]. Another biomass used to produce activated carbon is the garlic skin activated with KOH, this material presents a surface area of 2818 $m^2\ g^{-1}$, exhibiting excellent electrochemical performance and cycle stability at a current density of 0.5 $A\ g^{-1}$ (specific capacitance of 427 $F\ g^{-1}$ or 162 $F\ cm^{-3}$) and a retention capacitance of 94% [17].

Remarkably, the use of H_3PO_4 as activator leads to a lower surface area in comparison to those obtained using KOH or $ZnCl_2$. It implies a controlled porosity in the mesopore range. The addition of H_3PO_4 activator also enables to obtain surfaces composed of different functional groups such as quinones and phosphide groups -C-O-P-, which subsequently decompose into CO at 860°C. In addition, with this activation type, large capacitance values can be obtained, all due to the nature of the phosphorus functional groups in the carbon structure [23].

Another source of biomass as coconut was also activated using NaOH. In this case, Sesuk et al. showed that the carbon material presented a surface area of 2056 $m^2\ g^{-1}$ and specific capacitance of 192 $F\ g^{-1}$ at 1.0 $A\ g^{-1}$ [29].

Orange peels in copper carbonate ($CuCO_3$) were also used to produce activated carbon. These materials present a surface area of 912 $m^2\ g^{-1}$ and specific capacitance of 375 $F\ g^{-1}$ (current density of 1.0 $A\ g^{-1}$) [22].

Among the great variety of surface functional groups able to be incorporated on the material surface we can mention: OH^-, $COOH^-$, CO, as well as adsorbed molecules [25]. These functional groups increase their electron affinity in the aqueous medium, inducing electrochemical reactions [26, 30], which shows that their specific capacitance could be improved by addition of a pseudocapacitive component due to reversible faradaic redox involved in this type of molecules. In **Figure 2**, the optimized structures of 2,7-dimehtyl-9,10-phenanthrenequinone and its reduced form, adsorbed on a graphene surface obtained by density functional

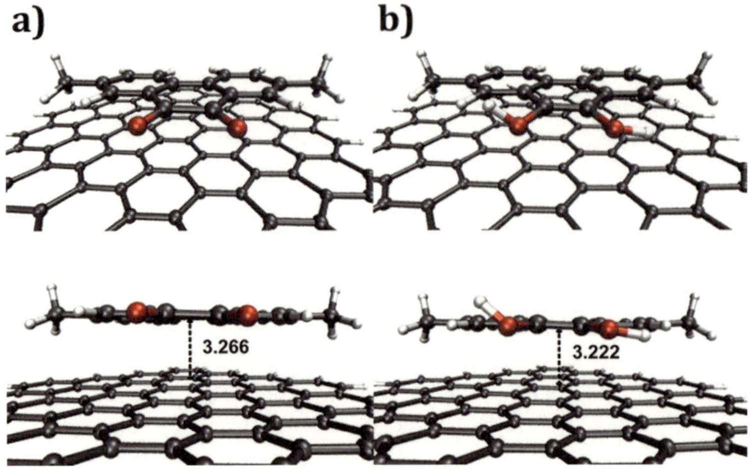

Figure 2.
Optimized structures of 2,7-dimehtyl-9,10-phenanthrenequinone (a) and its reduced form (b) adsorbed on a graphene surface, at the M06-2X/6-311G(d,p) level of theory. Distances in Angstroms.

theory (DFT) calculations at the M06-2X/6-311G(d,p) level of theory, are presented.

2. Reduced graphene oxide modified carbon Fibers

In the search for a low-cost electrode, with large surface area and optimal charge retention capacity, it has been found that carbon fiber or cotton exhibits adequate surface area properties and mechanical and electrochemical stability, in comparison with other porous carbon-based electrodes [31]. Certainly, different authors have reported that the micro and macro-porosity are formed during the lignite carbonization on cotton, but the tubular structure of the cellulose fibers is not altered in the pyrolysis process [32]. In this way, carbon cotton electrodes present a large surface area related to its micro and macro-porosity [31]. Another advantage linked to the production of these electrodes is the low cost and ease synthesis, *e.g.* Sheng-Heng et al. have reported commercial cotton carbonization at 900°C under Ar atmosphere for 6 h, with area of 805 $m^2 g^{-1}$ and average micropore size of 1 nm [33]. On the other hand, H. Wang et al. have obtained activated cotton carbons from cotton pieces treated with solution of KOH-H_2O or KOH-urea-H_2O at 700°C for 2 h under N_2 atmosphere. These activated cotton carbons show 1286 $m^2 g^{-1}$ of surface area and high electrochemical stability after several charge–discharge cycles as cathode in Li-S batteries [31].

In previous studies, the influence of reduced graphene oxide (r-GO) on carbon cloths and its capacity to charge store and stability have been studied. The carbonization of cotton fibers impregnated with graphene oxide at 1000°C for 2 h shows a reversible charge–discharge behavior of 160 mA h g^{-1} after 100 cycles [34]. In contrast, the design of carbon fiber material doped with nitrogen reports an outstanding stability after 200 charge–discharge cycles at 1 A h g^{-1} [34–36]. To a better understanding of modified carbon cotton cloths with r-GO on electrochemical performance, carbon cotton cloths were synthetized with r-GO 3 mg mL^{-1}, and the effect of inert atmosphere was evaluated. To achieve this, commercial cotton fibers were impregnated with graphene oxide (GO) during 0, 15, 30 and 45 min, after dried at 60°C the samples were pyrolyzed at 800°C (3°C min^{-1}) for 30 min under

N_2 or Ar atmosphere to obtain N/CC and N/CC/$rGO_{5\text{-}45}$, or Ar/CC and Ar/CC/$rGO_{5\text{-}45}$ electrodes, respectively (**Figure 3**).

According to the EDS analysis, cotton fibers (COT) report a carbon mass percent (% m of C) of 49.1%, which increases to 63.9% after the GO impregnation stage for COT/GO_{45}, presumably due to the presence of GO sheets on the cotton fibers surface. Similarly, after the heat treatment there is a carbon content increment of 95.6% for Ar/CC. This large increase has been associated with the CO bonds breaking during the cellulose fiber polymerization process when it is treated at temperature higher than 400°C [37], but without an apparent alteration of the fibrillary structure. The same behavior is observed for carbonized fibers covered by rGO, with values of 90.7% (N/CC/rGO_{45}) and 92.87% (Ar/CC/rGO_{45}) of carbon mass (% m of C), suggesting that the reduction of oxygenated groups in COT and the GO is higher under Ar atmosphere (**Table 1**). This implies that the composition, crystallinity, and porosity of carbonaceous electrodes are dependent on the atmosphere used during carbonization. Both N_2-made and Ar-made electrodes have a similar fibrillary structure as seen in **Figure 4a** and **c**. However, at high magnification, it is observed that samples N/CC/$rGO_{5\text{-}45}$ and Ar/CC/$rGO_{5\text{-}45}$ present a rough surface covered by a porous carbon layer, which would indicate that the rGO sheets were inserted efficiently and independently of the gas used in the heat treatment (**Figure 4b** and **d**). There are marked differences between both systems where samples treated under N_2 atmosphere present a more compact rough surface with a rather scaly appearance (**Figure 4b**). On the other hand, those heat treated under Ar atmosphere show layers of rGO with a laminar appearance around the fiber and regions of homogeneous roughness (**Figure 4d**).

IR spectroscopy permits to elucidate the type and degree of GO-COT, and rGO-carbon fibers (CC) interactions. The main vibrational modes corresponding to the

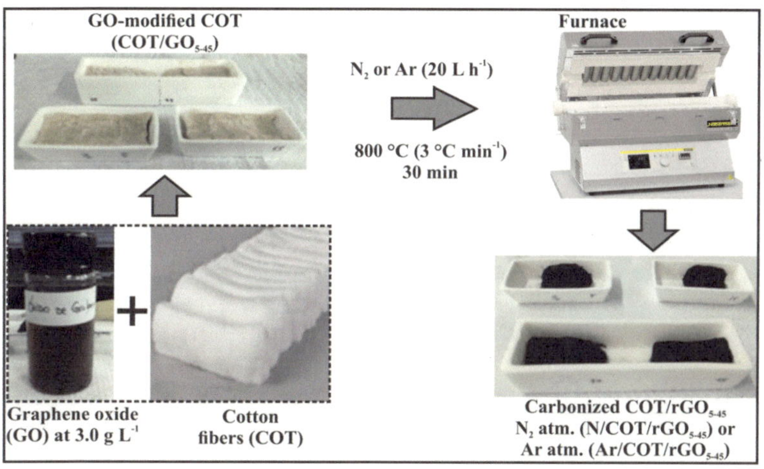

Figure 3.
Schematic diagram of the synthesis of rGO-modified carbonized cotton fibers.

Sample	COT	COT/GO_{45}	N/CC	N/CC/rGO_{45}	Ar/CC	Ar/CC/rGO_{45}
C mass %	49.1	63.9	93.3	90.7	95.6	92.9
O mass %	50.9	36.1	6.7	9.3	4.4	7.1

Table 1.
Carbon and oxygen mass percent obtained from EDX analysis for COT, COT/GO_{45}, N/CC, Ar/CC and Ar/CC/rGO_{45} samples [38].

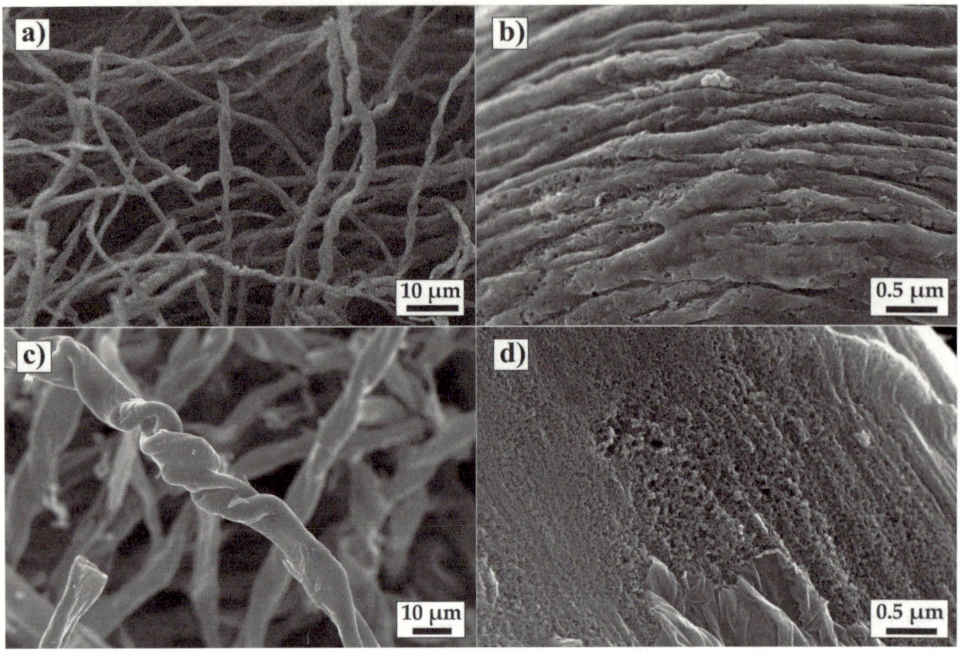

Figure 4.
SEM image of rGO-modified carbonized made fibers made under N_2 atmosphere (a, b) $N/CC/rGO_{45}$, and Ar atmosphere, (c, d) $Ar/CC/rGO_{5\text{-}45}$.

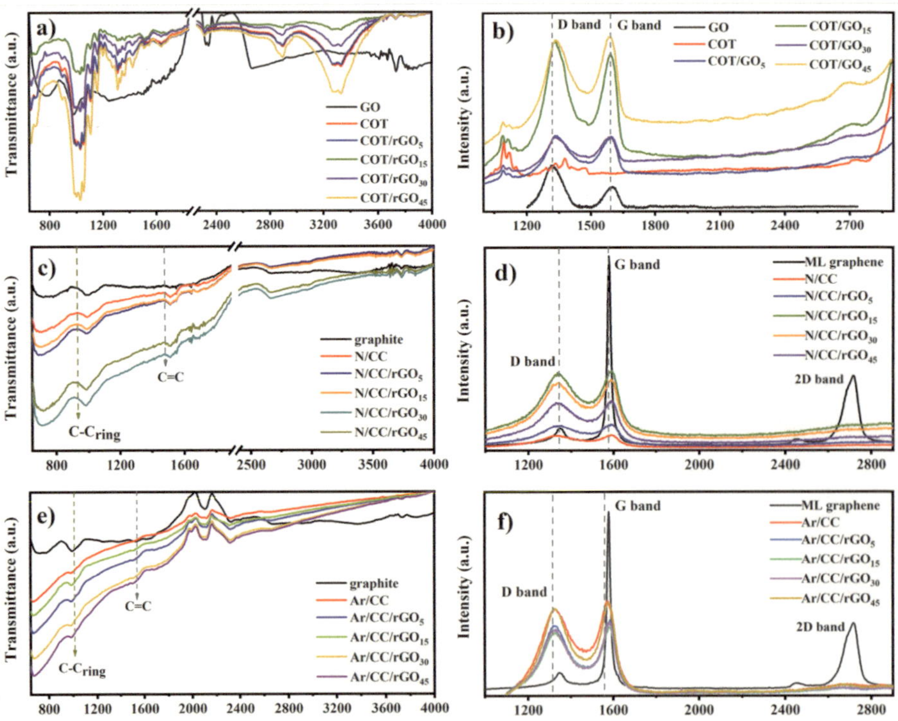

Figure 5.
Infrared and Raman spectra of (a, b) cotton fibers (COT) and GO-covered cotton fibers ($COT/GO_{5\text{-}45}$) and rGO-modified carbonized cotton fibers made under N_2 atm. (N/CC and $N/CC/rGO_{5\text{-}45}$) or Ar atm. (Ar/CC and $Ar/CC/rGO_{5\text{-}45}$). Raman laser source excitation of 532.5 nm.

characteristic chemical bonds present in COT are reported before and after the impregnation (COT and COT/$GO_{5\text{-}45}$) and the carbonization stage (N/CC/$rGO_{5\text{-}45}$ and Ar/CC/$rGO_{5\text{-}45}$), respectively (**Figure 5**). After the first stage, it is observed that the chemical groups C=O (1033.8 cm^{-1}), CH_2 (1322.2 cm^{-1}), C=O (1639.5 cm^{-1}), C-H (2893.2 cm^{-1}) and O-H (cellulosic, 3325.3 cm^{-1}) present an intensity reduction which could be correlated with GO layer. This reduction can also be associated with the interaction between the functional groups of GO and cellulose fibers. Since the cellulose OH and C=O groups register a decreasing in their intensities, it can be assumed that there exists a direct interaction between these chemical groups by hydrogen bonds [38]. According to IR spectroscopy, the bands at 979.8 and 1512.2 cm^{-1} are related to C-C_{ring} and C=C bonds, respectively, verifying a certain degree of graphitization. This thermal conversion has been described by M. M. Tang et al. who found that cellulose in cotton fibers undergoes a breakdown of the glycosidic bonds at temperatures above 250°C, allowing the elimination of various CO bonds and the breakdown of several C-C bonds along with a substantial mass loss due to H_2O, CO and CO_2 removal [39]. When the temperature is greater than 400°C, the remaining fractions of C-H hydrocarbon rings start an aromatization stage with H_2 loss and the formation of C=C bonds as part of a carbon polymeric structure [37]. It is observed that the conversion process to a graphitic like structure is similar for N/CC/$rGO_{5\text{-}45}$ and Ar/CC/$rGO_{5\text{-}45}$ (**Figure 5**).

The analysis of Raman spectra from **Figure 5b, d** and **f** show that N/CC/$rGO_{5\text{-}45}$ electrodes present a slight improvement in their crystallinity with respect to the modified cotton fibers, reporting a crystallite size from 19.39 (COT/GO_{30}) to 20.10 nm (N/CC/rGO_{30}), that is, an improvement of 4 to 5% (**Figure 6**). This behavior is also observed in the variation of the defect density from Ar/CC to Ar/CC/$rGO_{5\text{-}45}$ and from N/CC to N/CC/$rGO_{5\text{-}45}$.

For instance, Ar/CC/rGO_{30} register 21.9×10^{10} defects cm^{-2} compared to N/CC/rGO_{30}, whose value was 21.1×10^{10} defects cm^{-2} (**Figure 6**). According to Yo-Rhin Rhim et al. the peaks centered at 1620 cm^{-1}, between 1500 to 1550 cm^{-1} and about 1100 cm^{-1}, are associated with ordered π bonds, sp^2 amorphous systems,

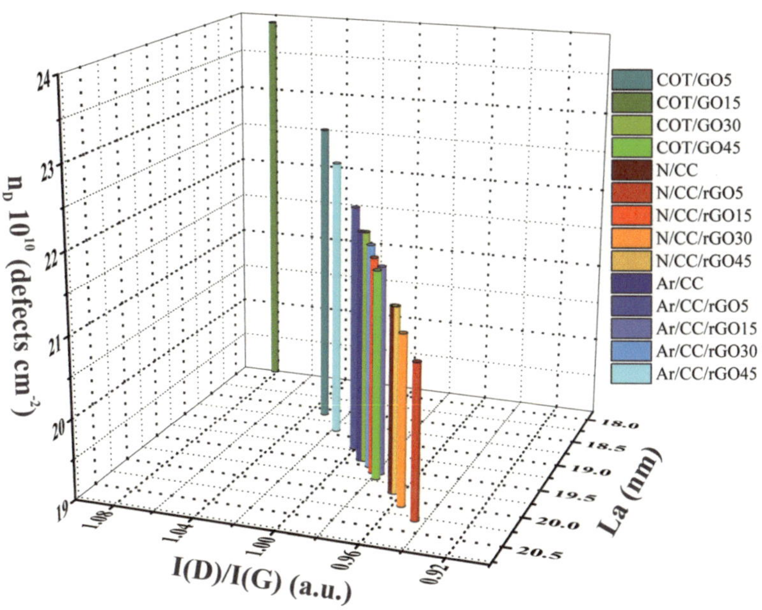

Figure 6.
3D graph of defect density (n_D) as function of I(D)/I(G) ratio and crystallite size (L_a) from Raman spectroscopy analysis.

and sp^3 bonds, respectively. Additionally, the systems under study were treated at 800°C, the restoration of the C=C bonds in the RGO is guaranteed and remains stable inside of the carbon fiber matrix [32]. Similarly, Ar/CC and Ar/CC/rGO_{5-45} electrodes show a slight increase in crystallite size, for example, Ar/CC/rGO_{30} registered a value of 19.5 nm in contrast to that reported for its precursor COT/GO_{30}, i.e., 19.39 nm (**Figure 6**). Although this variation is small, the effect of the carbonization stage is better appreciated when defect density is compared, e.g., Ar/CC/rGO_{30} shows 21.8×10^{10} defects cm^{-2} while COT/GO_{30} was 21.9×10^{10} defects cm^{-2}. This result means that the GO sheets supported on the cotton fibers present an excess of 0.1×10^{10} defects cm^{-2} compared to Ar/CC/rGO_{30} sample. It can thus be suggested that the nature of the inert atmosphere influences the degree of graphitization of the obtained products. Under N_2 atmosphere, a higher crystallinity is observed in N/CC/rGO_5 with a value equal to 20.3 nm and a defect density of 20.9 $\times 10^{10}$ defects cm^{-2}. In the case of cotton fibers modified with GO treated under Ar atmosphere, the optimal registered support is Ar/CC/rGO_{15} with a crystallite size of 19.6 nm and a defect density of 21.6 $\times$ 10^{10} defects cm^{-2}.

According to IR and Raman analysis, it is confirmed that GO sheets are intimately impregnated on cotton fibers through O-H or C=O interactions, and this interaction intensifies after the carbonization stage [38]. This effect is corroborated as an increase in crystallinity and a decrease in the density of defects (**Figure 6**). Therefore, through Raman spectroscopy it is possible to elucidate those carbonaceous materials that present a higher degree of crystallinity, lower defect density and higher density of sp^2 carbons which providing a high electrical conductivity. The evidence from these results suggests that samples such as N/CC/rGO_5, N/CC/rGO_{30}, Ar/CC/rGO_5 and Ar/CC/rGO_{30} are shown as promising matrices in the design of supercapacitors.

On the other hand, N_2 adsorption–desorption experiments were used to investigate the surface characteristics of the synthesized samples. It was possible to determine the specific surface area (S_{BET}), the degree and porosity type of the designed materials. In the case of N/CC presents a S_{BET} of 453.3 m^2 g^{-1}, shaped of a microporous area (S_{micro}) of 265.9 m^2 g^{-1} and with a size pore of 1.99 nm (**Table 2**). Furthermore, N/CC/rGO_{15} and N/CC/rGO_{30} exhibit S_{BET} values of 1221.3 and 1804.8 m^2 g^{-1}, respectively. These values suggest the presence of a highly porous rGO layer on carbonized cotton fibers prepared under N_2 atmosphere. Specifically, a high microporosity of 882.4 and 1350.2 m^2 g^{-1} is reported for the N/CC/rGO_{15} and N/CC/rGO_{30}, respectively, as a consequence of a large volume of micropores with diameters close to 1.88 nm (N/CC/rGO_{15}) and 1.80 nm (N/CC/rGO_{30}). Similarly, there was an increase in the S_{BET} from 1073.6 to 1457.5 m^2 g^{-1}, but with a substantial increase in mesoporosity from 298.6 to 729.8 m^2 g^{-1} for the Ar/CC and Ar/CC/

Samples		**S_{BET}/ m^2 g^{-1}**	**S_{micro}/ m^2 g^{-1}**	**S_{meso}/ m^2 g^{-1}**	**S_{micro}/S_{meso}**	**Pore/nm**
N_2-made electrodes	N/CC	453	266	187	1.4	2.0
	N/CC/rGO_{15}	1221	882	339	2.6	1.9
	N/CC/rGO_{30}	1805	1350	455	2.9	1.8
Ar-made electrodes	Ar/CC	1074	775	299	2.6	2.1
	Ar/CC/rGO_{15}	1458	728	730	1.0	2.0
	Ar/CC/rGO_{30}	1207	784	423	1.9	2.0

Table 2.
Surface properties registered by N_2 adsorption–desorption experiment [38].

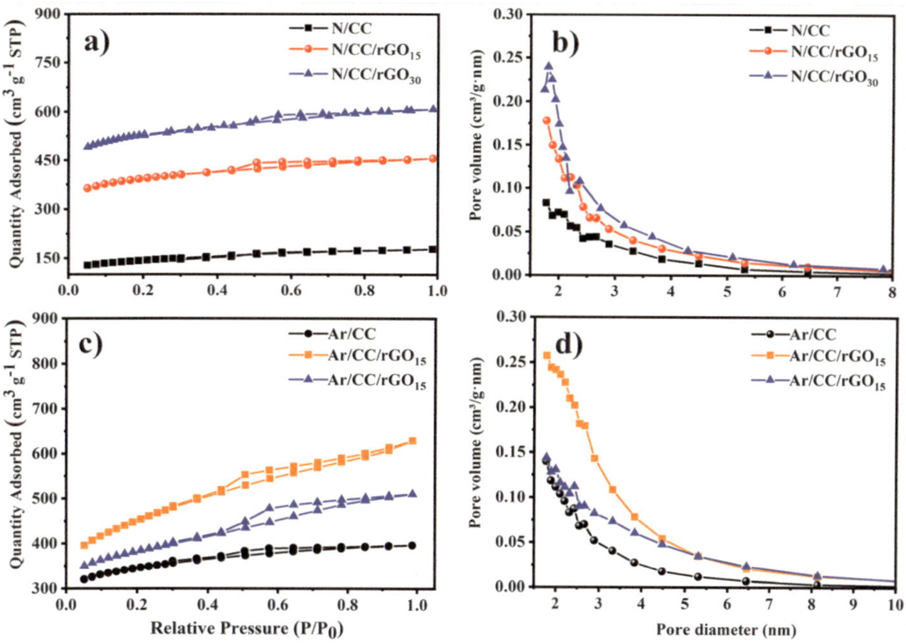

Figure 7.
N_2 adsorption–desorption isotherms and pore size distribution of carbonized cotton fibers made under (a, b) N_2 atmosphere. (N/CC and N/CC/rGO$_{5-45}$) and (c, d) Ar atmosphere. (Ar/CC and Ar/CC/rGO$_{5-45}$).

rGO$_{15}$, respectively **(Figure 7c)**. Additionally, it was observed that the Ar/CC/rGO$_{15}$ presents a balance between micro and mesoporosity with a ratio of 1: 1, in contrast to Ar/CC (2.6: 1) and Ar/CC/rGO$_{30}$ (1.9: 1). It can therefore be asserted that the electrochemical properties could be modified in the same way and correlated with the aforementioned texture properties [10].

Regarding to cyclic voltammetry test, a layer capacitive profile for N/CC (**Figure 8a**), Ar/CC, Ar/CC/rGO$_{15}$ and Ar/CC/rGO$_{30}$ (**Figure 8b**) is observed, while an increment of resistivity for N/CC/rGO$_{15}$ and N/CC/rGO$_{30}$ is registered

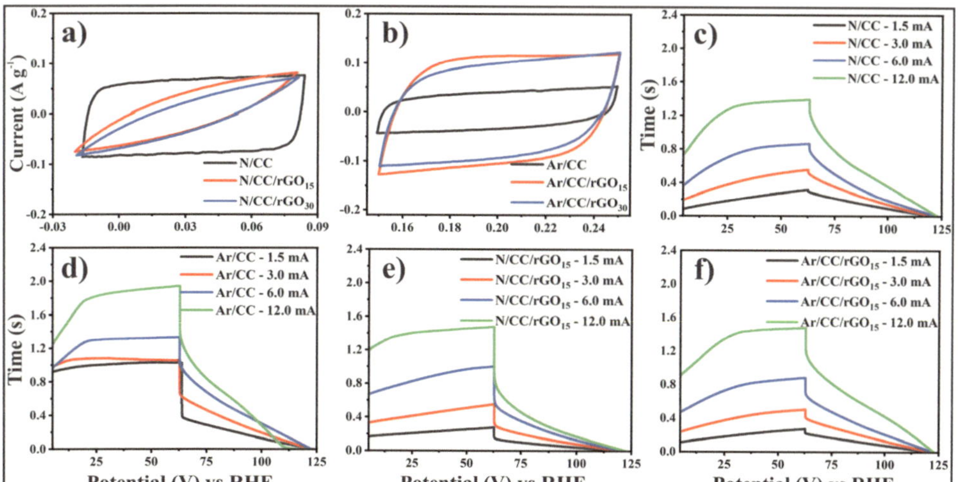

Figure 8.
Cyclic voltammetry of (a) N/CC, N/CC/rGO$_{15}$ and (b) Ar/CC, Ar/CC/rGO$_{15}$ and Ar/CC/rGO$_{30}$ at 1.0 mV s^{-1} in H_2SO_4 1.0 Mol L^{-1}. Charge–discharge test of (c) N/CC, (d) Ar/CC, (e) N/CC/rGO$_{15}$ and (f) Ar/CC/rGO$_{15}$ at 1.5, 3.0, 6.0 and 12.0 mA for 120 s.

(**Figure 8a**) [10]. This fact represents a reduction of N_2-made electrodes specific capacitance (C_s) from 70.3 (N/CC) to 45.3 F g^{-1} (N/CC/rGO_{15}), revealing a growing capacitive current for Ar-made electrodes at same potential range from 69.6 (Ar/CC) to 197.8 F g^{-1} (Ar/CC/rGO_{15}), as maximum. Both tendencies can be associated to the micro/mesoporosity ratio (S_{micro}/S_{meso}) showed in **Table 2**, owing to the high pore free energy that restricts the ionic charge transfer in the electrode-electrolyte interface [34, 40]. Thus, a high microporosity implies a limited ionic polarization at non-faradaic conditions, and a low capacitive charge retention [41]. In addition, capacitive current can be increased if the mesoporous surface area is extended, as it has been seen for Ar-made electrodes.

According to galvanostatic charge–discharge test (GCD) [42, 43], specific capacitance (C_{GCD}) of N/CC tends to decrease while the impregnation time increases (**Figure 8c**). This fact can be associated to the microporous surface area (S_{micro}) increase, as shown in **Figure 8c** and **e**. This can also be attributed to the presence of series-resistance in the pores (IRS) [34, 40]. Particularly, C_{GCD} decreases from 178 to 162 F g^{-1} at 0.4 A g^{-1} of applied current density for N/CC and N/CC/rGO_{45}, respectively (**Table 3**). On the other hand, if the inert atmosphere is replaced by Ar flux at the same thermal treatment conditions, Ar/CC shows a capacitive current increment with the impregnation time (**Figure 8d** and **f**). For instance, Ar/CC shows a C_{GCD} of 129 F g^{-1}, while its rGO-modified electrode, Ar/CC/rGO_{15}, reports a value of 219 F g^{-1} at 0.4 A g^{-1} (**Table 3**).

A detailed analysis of the electrochemical behavior by electrochemical impedance spectroscopy (EIS) experiments and non-linear complex fitting (NLCF) of their equivalent circuits has been conducted. Nyquist diagrams show an electrolyte resistance (R_S) of 2 Ω approximately, as well as, non-ideal impedance loop for both N_2-made and Ar-made electrodes suggesting a non-ideal charge storage process at the electrochemical interface [44, 45].

Regarding to N_2-made electrodes, charge transport resistance (R_1) increases with the GO impregnation time at high frequencies range (10^4–10^2 Hz) (**Figure 9a**). NLCF shows that R_1value varies from 5.6 (N/CC) to 100.3 Ω (N/CC/rGO_{30}), as maximum. Moreover, at medium frequencies range (10^2 – 10^{-1} Hz) N/CC electrode shows an *IRS* of 6.3 *Ω*, whereas N/CC/rGO_{15} reports values of 33.6 *Ω* and N/CC/rGO_{30} present an *IRS* conformed by two circuit elements, R_2 and R_3 whose values are 125.8 *Ω* and 149.3 *Ω*, respectively. These results suggest that ionic transport resistance in the inner porous surface is increased from N/CC/rGO_{15} to N/CC/rGO_{30} as their microporosity becomes higher [46–48].

For the Nyquist diagram (**Figure 9**) the equivalent circuit was calculated, resulting in ideal capacitive element (C_1) of 4.8 x 10^{-5} for N/CC/rGO_{30}. This result can be associated to the presence of the rGO layer on the fiber surface [49]. According to the transmission line model, N/CC/rGO_{30} internal capacitances are represented by constant phase elements (Q_2, Q_3), where their behavior is related to

N_2-made samples	C_{GCD}/F g^{-1}	IR_{drop}/V	Ar-made samples	C_{GCD}/F g^{-1}	IR_{drop}/V
N/CC	178	0.1	Ar/CC	129	0.5
N/CC/rGO_5	157	0.1	Ar/CC/rGO_5	185	0.3
N/CC/rGO_{15}	136	0.3	Ar/CC/rGO_{15}	219	0.2
N/CC/rGO_{30}	163	0.8	Ar/CC/rGO_{30}	202	0.1
N/CC/rGO_{45}	162	0.4	Ar/CC/rGO_{45}	165	0.4

Table 3.
Double layer electrochemical capacitance by galvanostatic charge–discharge test (C_{GCD}) for N_2- and Ar-made electrodes [38].

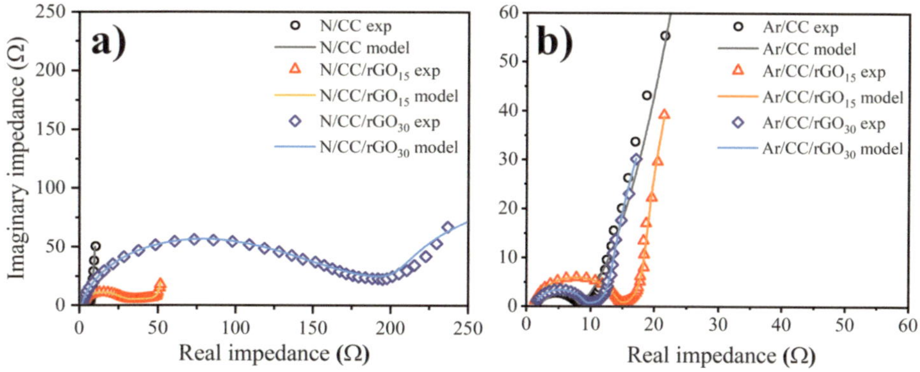

Figure 9.
Nyquist diagram of N_2-made electrodes (a) N/CC, N/CC/rGO$_{15}$ and N/CC/rGO$_{30}$ and Ar-made electrodes (b) Ar/CC, Ar/CC/rGO$_{15}$ and Ar/CC/rGO$_{30}$. Experimental (empty dots) and theoretical spectra (solid lines).

a non-ideal capacitor [44, 46]. Interestingly, the internal capacitance shows a minimum value of 0.2 $\Omega^{-1}\, s^{\alpha 2}$ (α_2, 0,90) for N/CC/rGO$_{30}$ (**Figure 10a**) [50]. Furthermore, N/CC/rGO$_5$ and N/CC/rGO$_{45}$ show a similar correlation with the impregnation time.

Besides, Ar-made samples report lower charge transport resistance (R_1) than N_2-made electrodes. For instance, Ar/CC shows a R_1 of 5.8 Ω, while Ar/CC/rGO$_{15}$ reports a value of 2.1 and Ar/CC/rGO$_{30}$ registers a R_1 of 2.8 Ω. As well, *IRS* is represented by a unique circuit element (R_2) of 12.8 and 7.3 Ω for Ar/CC/rGO$_{15}$ and Ar/CC/rGO$_{30}$, respectively. Remarkably, Ar-made samples report a laminar mesoporosity, suggesting that ionic diffusion is controlled by the pore characteristics. Hence, a finite diffusion element (M) is used to describe the capacitive behavior at low frequencies (**Figure 10b**) [46, 47, 50]. In this sense, Ar/CC/rGO$_{15}$ presents a constant phase element (Q_3) of 0.6 $\Omega^{-1}\, s^{\alpha 3}$ (α_3, 0.99), and a Warburg impedance (W) of 0.3 $\Omega^{-1}\, s^{0.5}$. On the other hand, Ar/CC/rGO$_{30}$ presents a non-

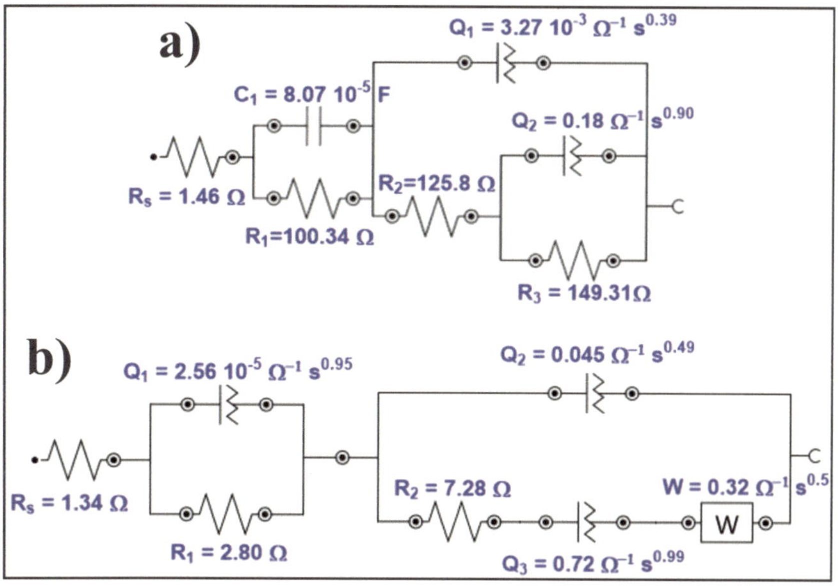

Figure 10.
Equivalent circuit of (a) N/CC/rGO$_{30}$, (b)Ar/CC/rGO$_{30}$. The inserted values are calculated from non-linear complex fitting of the EIS measurements.

ideal capacitance of Q_3 of 0.7 $\Omega^{-1}\, s^{\alpha 3}$ (α_3, 0.99) together to diffusion impedance W of 0.3 $\Omega^{-1}\, s^{0.5}$ (**Figure 10b**). In contrast, Ar/CC only shows an ideal capacitive behavior of 3.0 x 10^{-5} F (C_{dl}) and internal one of 0.2 F.

Accordingly, the total capacity (C_{EIS}) from the internal capacitive element (Q^{α}_{EIS}) is reported where the non-ideal constant (α) tends to one. For the estimation of the electrochemically accessible surface (S_{EIS}), a double layer charge density (Q^{o}_{dl}) of glassy carbon electrode of 3.0 x 10^{-5} F cm^{-2} is considered [51, 52],

$$S_{EIS}\left(m^2 g^{-1}\right) = Q_{EIS}^{\alpha\approx 1.0} / \left(mxQ^{o}_{dl}\right) \tag{6}$$

Additionally, the electrochemically accessible surface ratio (R_{ESA}) registers the fraction of interface surfaces available in the total physical surface area (S_{BET}) [52].

$$R_{ESA} = S_{EIS}/S_{BET} \tag{7}$$

The difference between C_{EIS} and C_{GCD} values show the total charge related to the electrochemical double layer, whereas the GCD test also registers the faradaic charge caused by the surface carbon and oxygen groups [52, 53]. As we can observe in **Table 4**, N_2-made samples present a decrease of the R_{ESA} (from 0.6 to 0.2) which can be related to the increment of the free energy adsorption promoted by their microporosity. The opposite occurs in Ar-made electrodes where R_{ESA} increases (from 0.2 to 0.4) and it can be related to their laminar mesoporous surface which results in a controlled ionic diffusion. Based on the R_{ESA} results, Ar/CC/rGO$_{15}$ and Ar/CC/rGO$_{30}$ electrodes emerge as promising candidates for the design of supercapacitors.

The aforementioned results enable us to conclude that the inert atmosphere has a strong influence on the surface and electrochemical characteristics of the synthetized carbon-based electrodes. Under N_2 atmosphere, N/CC and N/CC/rGO$_{5\text{-}45}$ show a remarkable microporous surface area. Unfortunately, this high microporosity affects the ionic diffusion, capacitive behavior and resistive character. Otherwise, under Ar atmosphere, an increase of the mesoporous surface is reported, based on a laminar pore distribution, associated with controlled ionic diffusion in the electrochemical cell, resulting in an increase of the capacitance.

Samples		C_{GCD}/ F g^{-1}	$Q_{EIS}^{\alpha\approx 1.0}$/ F	C_{EIS}/ F g^{-1}	S_{EIS}/ m^2 g^{-1}	S_{BET}/ m^2 g^{-1}	R_{ESA}
N_2-made electrodes	N/CC	178	0.3	85	285	453	0.6
	N/CC/rGO$_5$	157	0.2	60	198	—	—
	N/CC/rGO$_{15}$	136	1.0	112	373	1221	0.3
	N/CC/rGO$_{30}$	163	0.2	94	315	1805	0.2
	N/CC/rGO$_{45}$	162	0.3	57	191	—	—
Ar-made electrodes	Ar/CC	129	0.2	64	212	1074	0.2
	Ar/CC/rGO$_5$	185	0.4	108	359	—	—
	Ar/CC/rGO$_{15}$	219	0.6	172	573	1458	0.4
	Ar/CC/rGO$_{30}$	202	0.7	149	497	1207	0.4
	Ar/CC/rGO$_{45}$	165	0.3	111	369	—	—

Table 4.
Comparative chart between double layer electrochemical performance and superficial characteristics for N_2- and Ar-made carbon-based electrodes [38].

Remarkably, both Ar/CC/rGO_{15} and Ar/CC/rGO_{30} exhibit a promising capacitive behavior, as well as an optimal electrochemical accessible surface.

In this sense, the design of carbon-based electrodes from biomass-derived materials represents an outstanding way to obtain several energy-store electrochemical systems [54–56]. In addition, a remarkable capacity retention depending on pore size distribution looks like a constant effect on the constitution of double layer interface (DLI).

For instance, L. Jiang et al., reported a high performance of cellulose-derived microporous electrodes of 115 F g^{-1} and > 87% of capacity retention [54]. As well, Y. Zou et al., described rGO-intercalated carbon cloth fibers electrodes with a meso/micro-porous distribution and 64.5 mF cm^{-2} of specific capacitance [55]. Sheng-Heng C. et al. show that macro/micro-porosity to carbonized cotton fibers produce carbons with notable surface area [56]. On the other hand, Hui Shao et al., have paid special attention to nanoscale pores on carbon materials obtained by templates and double layer capacitance behavior at inside of this nanoscale pores [57]. Therefore, this fact is still discussed, owing to non-clear description is achieved by a cylindrical pore model in overall cases. Other parameters as the kind of inert atmosphere are considered in this work. Herein, a brief explanation of porosity-electrochemical performance is described, and inert atmosphere influence has been exposed as a significant parameter, as well as gas flux and temperature, certainly [38, 39].

Acknowledgements

The authors wish to thank the Vice-Rectorate of Research of National University of Engineering (UNI), Contract P-CC-2021-000546, and the Peruvian government agencies CONCYTEC and FONDECYT /World Bank (ref. 026-2019 FONDECYT-BM-INC.INV).

Conflict of interest

The authors declare no ‘conflict of interest’.

Author details

Antony Bazan-Aguilar[1], Elvis O. López[2], Miguel Ponce-Vargas[3] and Angélica M. Baena-Moncada[1*]

1 Research Laboratory in Applied Electrochemical, National University of Engineering (UNI), Lima, Peru

2 Department of Experimental Low Energy Physics, Brazilian Center for Research in Physics (CBPF), Rio de Janeiro, Brazil

3 Institut de Chimie Moléculaire de Reims UMR CNRS 7312, Université de Reims Champagne-Ardenne, Reims, France

*Address all correspondence to: abaenam@uni.edu.pe

References

[1] Nejat P, Jomehzadeh F, Taheri MM, Gohari M, Abd. Majid MZ. A global review of energy consumption, CO2 emissions and policy in the residential sector (with an overview of the top ten CO2 emitting countries). Renew Sustain Energy Rev. 2015;43:843–62.

[2] Oncel SS. Green energy engineering: Opening a green way for the future. J Clean Prod. 2017;142:3095–100.

[3] Viviescas C, Lima L, Diuana FA, Vasquez E, Ludovique C, Silva GN, et al. Contribution of Variable Renewable Energy to increase energy security in Latin America: Complementarity and climate change impacts on wind and solar resources. Renew Sustain Energy Rev. 2019 Oct;113:109232.

[4] Gao Z, Zhang Y, Song N, Li X. Biomass-derived renewable carbon materials for electrochemical energy storage. Mater Res Lett. 2017 Mar;5(2): 69–88.

[5] Meng Q, Cai K, Chen Y, Chen L. Research progress on conducting polymer based supercapacitor electrode materials. Nano Energy. 2017;36 (February):268–85.

[6] Wang J, Zhang X, Li Z, Ma Y, Ma L. Recent progress of biomass-derived carbon materials for supercapacitors. J Power Sources. 2020;451(January): 227794.

[7] Han X, Lu L, Zheng Y, Feng X, Li Z, Li J, et al. A review on the key issues of the lithium ion battery degradation among the whole life cycle. eTransportation. 2019 Aug;1:100005.

[8] Wentker M, Greenwood M, Asaba MC, Leker J. A raw material criticality and environmental impact assessment of state-of-the-art and post-lithium-ion cathode technologies. J Energy Storage. 2019 Dec;26:101022.

[9] Afif A, Rahman SM, Tasfiah Azad A, Zaini J, Islan MA, Azad AK. Advanced materials and technologies for hybrid supercapacitors for energy storage – A review. J Energy Storage. 2019;25(July): 100852.

[10] Gogotsi Y, Penner RM. Energy Storage in Nanomaterials - Capacitive, Pseudocapacitive, or Battery-like? ACS Nano. 2018;12(3):2081–3.

[11] Wang Y, Lei Y, Li J, Gu L, Yuan H, Xiao D. Synthesis of 3D-nanonet hollow structured Co3O4 for high capacity supercapacitor. ACS Appl Mater Interfaces. 2014;6(9):6739–47.

[12] Muzaffar A, Ahamed MB, Deshmukh K, Thirumalai J. A review on recent advances in hybrid supercapacitors: Design, fabrication and applications. Renew Sustain Energy Rev. 2019;101(October 2018):123–45.

[13] Frackowiak E, Abbas Q, Béguin F. Carbon/carbon supercapacitors. J Energy Chem. 2013;22(2):226–40.

[14] Li X, Wei B. Supercapacitors based on nanostructured carbon. Nano Energy. 2013;2(2):159–73.

[15] Simon Patrice, Brousse Thierry FF. Supercapacitors Based on Carbon or Pseudocapacitive Materials. 2017. 1–122 p.

[16] Zheng C, Qian W, Cui C, Xu G, Zhao M, Tian G, et al. Carbon nanotubes for supercapacitors: Consideration of cost and chemical vapor deposition techniques. J Nat Gas Chem. 2012;21(3): 233–40.

[17] Zhang Qing RK. Synthesis of Garlic Skin-Derived 3D Hierarchical Porous Carbon for High-Performance Supercapacitors. Nanoscale. 2018;1–9.

[18] Zhang G, Chen Y, Chen Y, Guo H. Activated biomass carbon made from bamboo as electrode material for supercapacitors. Mater Res Bull. 2018; 102(2010):391–8.

[19] Kumagai S, Sato M, Tashima D. Electrical double-layer capacitance of micro-and mesoporous activated carbon prepared from rice husk and beet sugar. Electrochim Acta. 2013;114:617–26.

[20] Yadav N, Ritu R, Promila P, Hashmi SA. Hierarchical porous carbon derived from eucalyptus-bark as a sustainable electrode for high-performance solid-state supercapacitors. Sustain Energy Fuels. 2020;7–35.

[21] Xu X, Zhou J, Jiang L, Lubineau G, Chen Y, Wu X-F, et al. Porous core-shell carbon fibers derived from lignin and cellulose nanofibrils. Mater Lett. 2013; 109:175–8.

[22] Wan L, Chen D, Liu J, Zhang Y, Chen J, Du C, et al. Facile preparation of porous carbons derived from orange peel via basic copper carbonate activation for supercapacitors. J Alloys Compd. 2020;823:153747.

[23] Elmouwahidi A, Bailón-García E, Pérez-Cadenas AF, Carrasco-Marín F. Valorization of agricultural wood wastes as electrodes for electrochemical capacitors by chemical activation with H_3PO_4 and KOH. Wood Sci Technol. 2020;54(2):401–20.

[24] Yu M, Han Y, Li J, Wang L. CO2-activated porous carbon derived from cattail biomass for removal of malachite green dye and application as supercapacitors. Chem Eng J. 2017;317: 493–502.

[25] Veerakumar P, Sangili A, Manavalan S, Thanasekaran P, Lin K-C. Research Progress on Porous Carbon Supported Metal/Metal Oxide Nanomaterials for Supercapacitor Electrode Applications. Ind Eng Chem Res. 2020;

[26] Wang J, Kaskel S. KOH activation of carbon-based materials for energy storage. J Mater Chem. 2012;22(45): 23710–25.

[27] Barzegar F, Khaleed AA, Ugbo FU, Oyeniran KO, Momodu DY, Bello A, et al. Cycling and floating performance of symmetric supercapacitor derived from coconut shell biomass. AIP Adv. 2016;6(11).

[28] Yin L, Chen Y, Li D, Zhao X, Hou B, Cao B. 3-Dimensional hierarchical porous activated carbon derived from coconut fibers with high-rate performance for symmetric supercapacitors. Mater Des. 2016;111: 44–50.

[29] Sesuk T, Tammawat P, Jivaganont P, Somton K, Limthongkul P, Kobsiriphat W. Activated carbon derived from coconut coir pith as high performance supercapacitor electrode material. J Energy Storage. 2019;25 (June):100910.

[30] Andreas HA, Conway BE. Examination of the double-layer capacitance of an high specific-area C-cloth electrode as titrated from acidic to alkaline pHs. Electrochim Acta. 2006;51 (28):6510–20.

[31] Wang H, Chen Z, Liu HK, Guo Z. A facile synthesis approach to micro–macroporous carbon from cotton and its application in the lithium–sulfur battery. RSC Adv [Internet]. 2014;4 (110):65074–80. Available from: http://xlink.rsc.org/?DOI=C4RA12260G

[32] Rhim YR, Zhang D, Fairbrother DH, Wepasnick KA, Livi KJ, Bodnar RJ, et al. Changes in electrical and microstructural properties of microcrystalline cellulose as function of carbonization temperature. Carbon N Y [Internet]. 2010;48(4):1012–24.

Available from: http://dx.doi.org/10.1016/j.carbon.2009.11.020

[33] Chung SH, Chang CH, Manthiram A. A Carbon-Cotton Cathode with Ultrahigh-Loading Capability for Statically and Dynamically Stable Lithium-Sulfur Batteries. ACS Nano. 2016;10(11): 10462–70.

[34] Chen L, Ji T, Mu L, Zhu J. Cotton fabric derived hierarchically porous carbon and nitrogen doping for sustainable capacitor electrode. Carbon N Y [Internet]. 2017;111:839–48. Available from: http://dx.doi.org/10.1016/j.carbon.2016.10.054

[35] Fan L-Z, Chen T-T, Song W-L, Li X, Zhang S. High nitrogen-containing cotton derived 3D porous carbon frameworks for high-performance supercapacitors. Sci Rep [Internet]. 2015;5(August):15388. Available from: http://www.nature.com/articles/srep15388

[36] Zhang X, Huang X, Zhang X, Zhong B, Xia L, Liu J, et al. A facile method to prepare graphene-coat cotton and its application for lithium battery. J Solid State Electrochem. 2016;20(5): 1251–61.

[37] Tang M., Bacon R. Carbonization of cellulose fibers—I. Low temperature pyrolysis. Carbon N Y. 1964;2(3):211–20.

[38] Bazan-Aguilar A, Ponce-Vargas M, Caycho CL, La Rosa-Toro A, Baena-Moncada AM. Highly Porous Reduced Graphene Oxide-Coated Carbonized Cotton Fibers as Supercapacitor Electrodes. ACS Omega [Internet]. 2020 Dec 22;5(50):32149–59. Available from: https://doi.org/10.1021/acsomega.0c02370

[39] Bazan A. Estudio Espectroscópico y Morfológico en el Grafeno. Universidad Nacional de Ingenieria; 2016.

[40] Zou Y, Wang S. Interconnecting carbon fibers with the in-situ electrochemically exfoliated graphene as advanced binder-free electrode materials for flexible supercapacitor. Sci Rep [Internet]. 2015;5(July):1–7. Available from: http://dx.doi.org/10.1038/srep11792

[41] Wang Y, Song Y, Xia Y. Electrochemical capacitors: mechanism, materials, systems, characterization and applications. Chem Soc Rev [Internet]. 2016;45:5925–50. Available from: http://dx.doi.org/10.1039/C5CS00580A

[42] Jiang L, Nelson GW, Kim H, Sim IN, Han SO, Foord JS. Cellulose-Derived Supercapacitors from the Carbonisation of Filter Paper. ChemistryOpen. 2015;4(5):586–9.

[43] Abushrenta N, Wu X, Wang J, Liu J, Sun X. Hierarchical Co-based Porous Layered Double Hydroxide Arrays Derived via Alkali Etching for High-performance Supercapacitors. Sci Rep [Internet]. 2015;5(July):1–9. Available from: http://dx.doi.org/10.1038/srep13082

[44] Barsoukov E, Macdonald JR. Impedance Spectroscopy [Internet]. Impedance Spectroscopy: Theory, Experiment, and Applications. 2005. 1–595 p. Available from: http://doi.wiley.com/10.1002/0471716243

[45] Hasyim MR, Ma D, Rajagopalan R, Randall C. Prediction of Charge-Discharge and Impedance Characteristics of Electric Double-Layer Capacitors Using Porous Electrode Theory. J Electrochem Soc. 2017;164(13):A2899–913.

[46] Klink S. In-depth analysis of irreversible processes in lithium ion batteries. Ruhr-Universität Bochum, Universitätsbibliothek; 2013.

[47] Newman JS, Tobias CW. Theoretical Analysis of Current

Distribution in Porous Electrodes. J Electrochem Soc. 2007;109(12):1183.

[48] ZIVE-Lab. ZMAN 2.3.2 User's Manual. ZIVE-Lab W, editor. Seoul, Korea: WonATech Co., Ltd; 2014. 1–173 p.

[49] Yang C. Reduced Graphene Oxide–Based Microsupercapacitors. In: Jiang Z, Kyzas G, editors. Rijeka: InTech; 2017. p. Ch. 6. Available from: https://doi.org/10.5772/67433

[50] Kang J, Wen J, Jayaram SH, Yu A, Wang X. Development of an equivalent circuit model for electrochemical double layer capacitors (EDLCs) with distinct electrolytes. Electrochim Acta [Internet]. 2014;115:587–98. Available from: http://dx.doi.org/10.1016/j.electacta.2013.11.002

[51] Fernández PS, Castro EB, Real SG, Visintin A, Arenillas A, Calvo EG, et al. Electrochemical behavior and capacitance properties of carbon xerogel/multiwalled carbon nanotubes composites. J Solid State Electrochem. 2012;16(3):1067–76.

[52] Fernández PS, Arenillas A, Calvo EG, Menéndez JA, Martins ME. Carbon xerogels as electrochemical supercapacitors. Relation between impedance physicochemical parameters and electrochemical behaviour. Int J Hydrogen Energy. 2012;37(13): 10249–55.

[53] Martin R, Quintana JJ, Ramos A, Nuez I De. Modeling Electrochemical Double Layer Capacitor , from Classical to Fractional Impedance. 2008;61–6.

[54] Luyun Jiang, Geoffrey W. Nelson, Heeyeon Kim, I. N. Sim, Seong Ok Han, and John S. Foord. Cellulose-Derived Supercapacitors from the Carbonisation of Filter Paper. ChemistryOpen 2015, 4, 586–589. DOI: 10.1002/open.201500150.

[55] Yuqin Zou, Shuangyin Wang. Interconnecting Carbon Fibers with the In-situ Electrochemically Exfoliated Graphene as Advanced Binder-free Electrode Materials for Flexible Supercapacitor. Scientific RepoRts. 5: 11792. DOi: 10.1038/srep11792.

[56] Sheng-Heng Chung, Chi-Hao Chang, Arumugam Manthiram. A Carbon-Cotton Cathode with Ultrahigh Loading Capability for Statically and Dynamically Stable Lithium–Sulfur Batteries. ACS Nano 2016, 10, 11, 10462–10470

[57] Hui Shao, Yih-Chyng Wu, Zifeng Lin, Pierre-Louis Taberna, Patrice Simon. Nanoporous carbon for electrochemical capacitive energy storage. Chem. Soc. Rev., 2020, 49, 3005. DOI: 10.1039/d0cs00059k

Chapter 3

Biomass Based Materials in Electrochemical Supercapacitor Applications

Sema Aslan and Derya Bal Altuntaş

Abstract

Biomass is the general term for organic substances derived from living organisms (plants and animals). Since, biomass is a renewable, sustainable, innovative, low cost and carbon-neutral energy source, the applications of nano-micro particles produced from biomass in electrochemical applications have emerged. A large number of carbon-based materials, such as featured activated carbon, carbon nanotube, C-dots, biochar, hybrid carbon-metal/metal oxide ... etc. can be produced from divergent types of biomass. With the growing energy need in the world, supercapacitors have also developed considerably besides the energy generation and storage methods. The supercapacitor is an energy storage system that can work reversibly to provide high energy in a short time. In these systems, electrode structure and surface properties are crucial for energy capacity enhancement. In this sense, electrode modifications with the above-mentioned biomass-based nano-micro structures are widely used in supercapacitor applications.

Keywords: Biomass, carbon, energy, electrochemistry, supercapacitor

1. Introduction

Biomass is defined as all biomass of non-fossil organic matter of biological origin, which can be renewed in less than 100 years, includes land and water-grown plants, animal wastes, food industry, and forestry by-products, and urban wastes [1]. Energy obtained from biomass sources such as vegetable resources, agricultural and animal wastes, organic origin city, and industrial wastes is defined as biomass energy. Since the basis of biomass energy is based on the photosynthesis of plants, biomass energy can also be expressed as the energy of organic matter where solar energy is stored as chemical energy [2].

Vegetable or animal biomass energy sources that can be found in the sea and/or land can be queued as; wood (energy forests, wood residues), oilseed crops (sunflower, rapeseed, soy, safflower, cotton, etc. ...), carbo-hydrate plants (potato, wheat, corn, beet, etc. ...), fiber crops (flax, kenaf, hemp, sorghum, etc. ...), vegetable residues (branches, stalks, straw, roots, bark, etc. ...), animal wastes, urban and cleans [1–3]. Biomass resources are of high water and oxygen content, low density, low calorific value; These features negatively affect the quality of loss [4]. The negative properties of biomass can be eliminated by physical processes (size reduction-crushing and grinding, drying, filtration, aggregation) and transformation processes (biochemical and thermochemical processes) [5]. Biomass energy stands out with its various advantages such as the ability to be grown almost everywhere, good knowledge of production and conversion technologies, suitability for energy production every time, adequacy of low light intensities, being storable, adequacy of temperatures between 5 and 35°C, being in socioeconomic developments, not creating environmental pollution (Very low NO_x and SO_2 emissions), less greenhouse effect formation compared to other energy sources, cream of CO_2 balance in the atmosphere, not causing acid rain [1–6].

2. Nanomaterials

Studies on nanoscale materials have shown very great improvement and received much attention for decades. Structures are defined as nanoscale materials; Nanocrystals are divided into different classes such as nanoparticles, nanotubes, nanowires, nanorods, or nano-thin films. The main reason for the focus on this subject is that the substances exhibit unusual properties and functionality in a certain size range, unlike their volumetric structures [7, 8]. On the other hand, nanoparticles, which are defined as powders with a size of 100 nm or less, form the basis of nanotechnology due to nanosized materials. These particles show properties that are generally considered different and superior to other commercial materials. The reasons known today for the attractiveness of the frequently mentioned nanoparticle properties are; Quantum size effects stand out as the size dependence of the electronic structure, unique characters of surface atoms and high surface/volume ratio [9].

Nanoparticle synthesis has paved the way for the preparation of many technological and pharmacological products such as high-activity catalysts, special technological materials for optical applications, superconductors, anti-wear additives, surfactants, drug carriers and special diagnostic tools, due to the extraordinary properties these structures exhibit. In addition, the control of materials at the nanoscale level allows the realization of miniaturized devices with specific functions such as nanocarriers, sensors, nanomachines and high density data storage cells [8–10]. The indispensable first step for new developments in nanotechnology, which includes the design, manufacture and functional use of nanostructured materials and devices, is the production of nanoparticles. Nanoparticles, which form the starting point of nanotechnological materials, can be produced in a wide chemical range and morphology. Today, nanoparticles with different morphologies such as core-shell, doped, sandwich, hollow, spherical, rod-like and polyhedral can be prepared from metal, metal alloy, ceramic and polymer-based or mixtures with the desired properties [10–12].

2.1 Biomass based carbon nanomaterials

Biomass is a very low cost carbon source and is used in the production of high value carbon nanomaterials [1–3]. While biochar, activated carbon and mesoporous graphite come first among these materials, when advanced processes are applied, graphene, graphene oxide, carbon nanotubes, carbon-based nanostructures and even hybrid nanostructures containing metalloxide can be successfully produced (**Figure 1**) [13–18].

In particular, biochar, activated carbon and porous carbon production from biomass and their various applications were very popular [16–21]. In one of these studies biochar is synthesized from hair and then activated charcoal without any chemical treatment [19]. The same group reported the synthesis and characterization

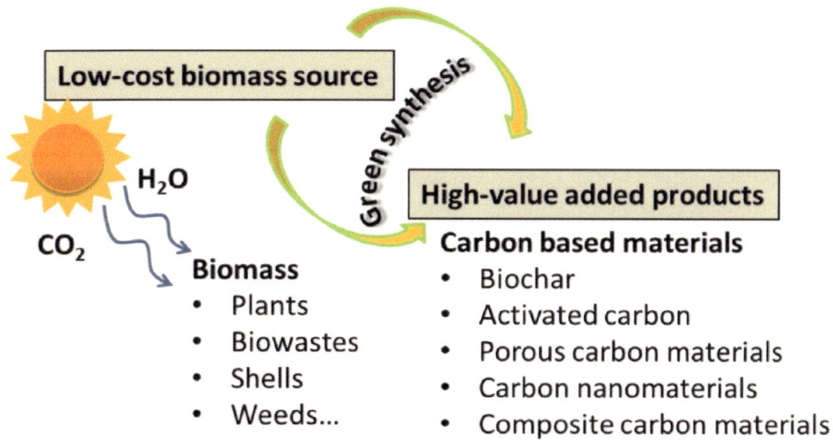

Figure 1.
General flowchart of the green synthesis of high value added carbon materials from the biomass.

of the activated carbon again from waste human hair mass successfully by using chemical activation [20]. Although biochar has a developed porous structure, it shows the limited specific surface area, and activated charcoal has an uncontrolled pore size distribution and irregular three-dimensional structure, which has greatly limited their use in specific applications. For this reason, new generation composite carbon structures originating from biomass were developed, and materials with superior thermal and electrical conductivity and good chemical and environmental stability were developed [22–24]. Later, graphene, graphene-like nano-layers were developed from structures such as activated charcoal obtained from biomass, whose pore structure progressed in a stable pattern with good electrical conductivity, and whose pore structure did not differ [16–22]. Also, Wang et al. [25] reported that the microwave plasma beam method can be utilized to create graphene CNT hybrids.

3. Supercapacitors

The need for energy is increasing day by day and the resources used for energy production are rapidly being depleted. Although the types of resources used in energy production vary, the dominant resources are still petroleum-based fossil fuels [26]. The greenhouse gases released while obtaining energy from these fuels threaten the ecosystem seriously. In this sense, energy storage has gained importance as well as energy production. The methods used for energy storage can be listed as ultracapacitors/supercapacitors, superconducting magnetic energy storage, fuel cells, and batteries [27]. Among these, supercapacitors have an important place. Supercapacitors are also known as secondary batteries used as energy storage and conversion systems. It has attracted great attention in recent years due to its high power density and long cycle life compared to rechargeable batteries [28]. Supercapacitors can be classified as electrical double layer capacitors, pseudocapacitors and hybrid capacitors. The most important factor in improving the performance of supercapacitors is the electrode used [27–29]. Many modifications have been developed to load on the electrode surface and to keep this charge in the double layer for the desired time, that is, to provide charging and discharging when needed. Conductive polymers [30], nanofibers [31], graphene and graphene oxide [32], carbon-based [19] or metal–metal oxide derivative hybrid [33] or nanostructures of different sizes and many other types of modifications.

3.1 Biomass based materials in supercapacitors

Carbon based materials occupy a respectable area on the electrode development in electrochemical energy devices. In general, the source and production methods of carbon-based materials used in batteries, capacitors, supercapacitors and fuel cells have gained importance [34]. The ability to obtain carbon-derived composite or single materials that exhibit superior properties from biomass has made an important contribution to this field. The materials produced to provide high capacitance, fast response and high cycle number, especially in supercapacitor applications [34, 35]. Especially the large surface area due to the porous structure of the carbon-based materials obtained and the ability to be made with an cost-effective and unlimited precursor provides a great advantage [34–42].

Senthil et al. [36] previously examined changes in supercapacitor performance by using the tubular porous carbon material (HT-PC) they produced using waste feather grass flower (FFGF) in electrode modification. They reported a very high specific capacitance value around 300 F g^{-1} with HT-PC containing electrode. Additionally, they obtained % 96 capacitance conservation after 50000 cycles whereas the electrolyte solution is changed from KOH to 1 M Et_4NBF_4/AN electrolyte they observed % 30 capacitance loss over 10000 cycles. This is noted as a very appreciatable recovery for capacitance conservation. In another flower based study is reported by Zheng et al. [37]. They used waste-kapok flower as a precursor to produce hierarchically porous carbon as supercapacitor electrode. Here the authors took the advantages of the oxygen rich structure of the kapok flower to obtain micro and meso porous carbon structure. The KOH electrolyte using supercapacitor achieved around 290 F g^{-1} supercapacitance value and showed excellent cyclic stability.

Activated carbons are one of the most popular carbon materials for commercial supercapacitors. Jain et al. [38] reported t a cavitation process of the activated carbon from the mixture of native European deciduous trees, Birch, Fagaceae, and *Carpinus betulus* (commonly known as European hornbeam). They reported a desirable carbon with enhanced porosity and high specific surface area of about 614 m^2 g^{-1}. Here they examined the supercapacitance of the synthesized activated carbon as an additive to electrode structure in acidic 1.0 M H_2SO_4 electrolyte. Finally, they showed that the reported method is suitable to achieve a versatile electrode material for supercapacitors.

The hydrothermal method is also a green synthesis method for biomass based productions. Nguyen et al. [39] utilized a carbonization method to synthesize a high-surface-area carbon (HSAC) material from the peanut shells. The HSAC growth on the Ni foam and obtained material achieved nearly two times higher specific surface area than the activated carbon which is produced by Jain et al. Besides they processed the supercapacitor application in KOH and 1 M Li_2SO_4 electrolyte and obtained a highly desirable capacitance value around 250 F g^{-1}. This is a very unique example of a completely green synthesis of carbon based materials in terms of the source and the processed method. Another impactive biomass based supercapacitance materials are carbon aerogels. They possess highly multifunctional features including compressibility and elasticity. Differently, Long et al. [40] used the calcinated mixture of glucose & dicyandiamide nanosheets (C-GD) and cellulose nanofibers (CNFs) to synthesize nitrogen doped carbon aerogels (C-NGD). C-GD and CNFs lead to a super stable wave-layered structure with an ultimate compression strain (95%) and dedicated as a potential multifunctional material toward flexible electronics, and energy conversion/storage devices.

The porosity of the developed carbon material is very crucial for catalytic applications. To achieve a better regularity in the pore size and distribution divergent methods are developed. This is very important to obtain consistent activities from the contributed material. In one of these studies, Wu et al. [41] reported that by using melamine foam it is possible to obtain hierarchical porous carbon. They used the waste liquid of the vitamin-C production as the precursor and obtained highly porous carbon by using this template. They used KOH activation on the precursor, and the porous material showed excellent supercapacitance value as 217 F g^{-1} in 1-ethyl-3-methylimidazolium tetrafluorobo- rate (EMIMBF 4) electrolyte. The cycling stability of the material amendable and the energy and power densities are evaluated as impactive by the authors (**Table 1**). A considerable part of this study is the utilization of the developed material in all-solid state state symmetric supercapacitors and obtained attainable capacitance value as 180 F g^{-1} capacitance in gel polymer electrolyte. Because, the multiple applications of the same material enhances their potential and commercial value. Also Fang et al. [42] developed a high-performance flexible supercapacitor which is produced from carbon nanorod and carbon fiber. They used waste straw for the production of carbon nanorods supported hydrothermal carbons and carbon fibers (CNR/HTC/CFs). They reached to 270 F g^{-1} in solid state supercapacitor.

It is possible to apply the produced featured biomass derived carbon materials in both HER and supercapacitors to evaluate their efficiency in energy devices. Cao et al. [52] used the bean sprout to produce nitrogen doped carbon material. Both the carbon and the nitrogen source was the biomass itself. In such a process the control of the pore size and distribution can be controlled by the heat treatment. Here the self-nitrogen doped porous structure evaluated as a good electrolyte ion transferring system for the hydrogen evolution reaction (HER) and supercapacitor applications. They reported very satisfactory HER and specific capacitances and showed that bean sprout have great energy potential for the industrial scale-up with low-cost.

A large amount of the biomass based production methods are covering lignocellulosic biomass residues. Some of the pioneering studies are reported by Selvaraj et al. and Tan et al. Selvaraj et al. [51] produced the activated carbon nanosheets from *Prosopis juliflora* wood carbon waste blocks as carbon precursor. They achieved an ultra-high specific surface area with micro/meso and macro pores and gravimetric capacitances from 400 to 430 F g^{-1} for the different supercapacitance measurement conditions. Tan et al. [8] reported excellent specific surface area and specific capacitance values (**Table 1**) by tuning the initiator biomass material and activation agent. They processed divergent precursors such as cellulose, hemicellulose, and lignin with changing percentages of extractives. It can be clearly seen that the specific surface area enhancement directly brings the increase in specific capacitance value. Yakaboylu et al. [53] reported that the tuning of the pretreatment parameters is very important to control the final pore composition in lignocellulose based activated carbon production. They used the Miscanthus grass biomass as the precursor for the sheet-like activated carbon synthesis and obtained around 190 F g^{-1} capacitance value. The pore size and distribution are controlled by the KOH pretreatment and interconnected micropores are obtained. Thus, the importance of the process control comes out which affects the pore size and volume, cellulose and oxygen amount in the structure, and morphological features which are crucial for the electrochemical adsorption capacity and catalytic performances of produced carbon material.

Sometimes the produced materials can be functionalized by additional groups or heat treatment is applied to enhance the electrical performance. The addition of heteroatoms provides interconnected networks between pores. Chaparro-Garnica et al. [54] synthesized highly porous (SBET >1200 m^2 g^{-1}) activated carbon from hemp residue by H_3PO_4-assisted hydrothermal carbonization (HTC). Then they doped the obtained activated carbon with nitrogen groups and stabilized the

Sample	Biomass sources	Activation	Synthesis method	Surface area (m^2/g)	Morphology	Capacitance	Electrolyte	Ref
C	Fig-fruit	KOH	Pre-oxidation and activation	2337	Highly porous foam-like structure	217 F g^{-1} at 20 A g^{-1} with three-electrode system	0.5 M H_2SO_4	[43]
N, S-C	Pomelo peel	KOH	Carbonization and activation	2206	Porous structure with micropores	317 F g^{-1} at 1.0 A g^{-1} with three-electrode system	1 M H_2SO_4	[44]
Co3O4/C	Mollusk shell	—	Carbonization and hydrothermal	—	Cubic Co_3O_4 (1 μm) coated on C (pore diameter about 25 μm)	1307 F g^{-1} at 1.0 A g^{-1} with three-electrode system	6 M KOH	[45]
NiCo2O4/C	Mollusk shell	—	Carbonization and hydrothermal	—	$NiCo_2O_4$ nanowires (1.5 μm) grew on Honeycomb-like C	1696 F g^{-1} at 1.0 A g^{-1} with three-electrode system	2 M KOH	[46]
Ni-Co/N-C	Bacterial cellulose	—	Carbonization and solution co-deposition	405.8	Ultrathin Ni-Co LDH nanosheetsuniformly anchored on C nanofibers	1949.5 F g^{-1} at 1 A g^{-1} with three-electrode system	6 M KOH	[47]
K-C	Banana stem	KOH	Dehydration, porogenic stage and pyrolysis	567	Non-graphitic carbon materials with highly disordered nano-crystalline structure of hard carbon	479.23 F g^{-1} 5 A g^{-1}	6M KOH	[48]
P-C	Banana stem	H_3PO_4	Dehydration, porogenic stage and pyrolysis	178	Non-graphitic carbon materials with highly disordered nano-crystalline structure of hard carbon	202.11 F g^{-1}	6M KOH	[48]
C	Corncobs	Pyrolysis	Direct pyrolysis	215	Non-graphitic carbon materials with highly disordered nano-crystalline structure of hard carbon	309.81 F g^{-1} 5 A g^{-1}	6M KOH	[48]
S-C	Potato starch	Pyrolysis	Preheating and carbonization	42	Non-graphitic carbon materials with highly disordered nano-crystalline structure of hard carbon	99.9 F g^{-1} 0.3 A g^{-1}	6M KOH	[48]

Sample	Biomass sources	Activation	Synthesis method	Surface area (m^2/g)	Morphology	Capacitance	Electrolyte	Ref
C	Walnut shells	KOH	Carbonization and activation	3577	A sheet-like activated carbon with thin-layer pore walls	220 F g^{-1} at 100 A g^{-1}	6 M KOH	[49]
C	Phoenix leaves	Heat and K_2FeO_4	Graphitization and activation	2208	Large micropores (more than 71.8%)	254 and 273 F g^{-1} at a current density of 0.5 A g^{-1}	KOH and H2SO4	[50]
C	Waste liquid from the production of vitamin C	KOH	Carbonization and activation	3837	Hierarchical porous carbon	217 F g^{-1} at 0.1 A g^{-1}	EMIMBF 4	[41]
C	Waste liquid from the production of vitamin C	KOH	Carbonization and activation	3837	Hierarchical porous carbon	180 F g^{-1} at 0.5 A g^{-1}	EMIMBF 4 / PVDF-HFP gel polymer electrolyte	[41]
C	*Prosopis juliflora* wood	Heat and KOH	Carbonization and activation	2943	Rational micro/meso/macro pore size distributed activated carbon nanosheets	588 F g^{-1} at 0.5 A g^{-1}	6 M KOH	[51]
C-K	Cellulose, hemicellulose, lignin	KOH	Carbonization and activation	3135	Hierarchical porous carbon	410.5 F g^{-1} at 0.5 A g^{-1}	6 M KOH	[8]
C-S	Human hair	—	Carbonization	—	Graphene like structure	139.00 F g^{-1}	6 M KOH	[19]

Table 1.
Properties and the supercapacitor performances of some of the biomass derived carbon based materials.

material with heating. Resulting highly porous activated carbon is applied to a supercapacitor and it showed comparable performances in aqueous and organic electrolytes. In another study, Liang et al. [55] reported nitrogen and sulfur co-doped hierarchical porous carbon (NSPC) with a high gravimetric capacitance around 350 F g^{-1}. They used the $NaHCO_3/KHCO_3$ activated foxtail grass seeds as precursor biomass. Also, Cao et al. [52] studied the effect of nitrogen containing groups of NH_4Cl, $(NH_4)_2CO_3$ and urea on the electrochemical performances of biomass derived hierarchical porous materials. They reported that, NH_4Cl is proved to be the porogen with the minimum collapse of pollen grain and urea can be identified as the most effective N dopant with the 300 F g^{-1} capacitance value. Du et al. [56] presented a silica activation process for the carbon produced from the carrot biomass. In this study nitrogen enriched porous carbon is produced by a simple activation method. Nitrogen enrichment is preferred to achieve a higher porous structure. Low-cost Na_2SiO_3 served as initiator and provided catalytic effect on the nitrogen doping process. In general the resulting material showed around 270 F g^{-1}. These results showed that the choice of the biomass precursor is very important that, while carrots are used as the carbon source, the vitamins in carrot biomass can serve as a nitrogen reserve. Ariharan et al. [57] reported a facile synthesis of self-phosphorous doped porous carbon material from Honeyvine milkweed (Pod fluff as a precursor) by a simple carbonization route without using any activation process under argon gas atmosphere. They achieved nearly 250 F g^{-1} capacitance value and showed an excellent supercapacitance recovery after 10000 cycles with 95% recovery. They also examined the H_2 storage capacity of the synthesized porous material and achieved successful performances. The reported study offers an effective route for the stable, conductive and highly porous carbon based material synthesis and their effective utilization in both energy applications.

Waste paper cups were used as a source for the synthesis of a carbon support that is loaded with Fluorescein molecules [58]. The resulting composite is successfully applied to a supercapacitor electrode and showed 214 F g-1 specific capacitance. Here the main point is that waste lignocellulosic materials are also a very promising tools for electrochemical response enhancement.

Mostly, the regions and their agricultural potentials are determinative for biomass source selection. In China rice based carbon production is very promising and their wastes are also bear huge biomass potentials. Xie et al. [59] used puffed-rice as a precursor material for carbon sheet structures. They applied gradual heat treatment to the source material and achieved the best performance from R-800 sample. They reached nearly 120 F g^{-1} capacitance performance. It is moderate but energy and power densities are recorded as attainable performances. They also utilized the material in microwave absorption beyond a supercapacitor material. So it has been shown that true selection of the biomass precursor can construct a bridge from the biomass derived materials to sustainable development. Orange peels are also used as a biomass source for activated carbon production [60]. The produced activated carbon is indicated as a very highly porous structure and its nanocomposite was produced by the combination with poly aniline. They both utilized for the supercapacitor material. Hybrid structure exhibited nearly 4 folds of higher capacitance value than the natural form. Another citrus based study is reported by Gehrke et al. [61]. The activated carbons synthesized from *Citrus bergamia* peels by activation with phosphoric acid (AC - H_3PO_4) and manganese nitrate (AC – Mn_3O_4). Among these materials AC – Mn_3O_4 exhibited the best electrochemical performance due to the active transition metal content with a specific capacitance value of 290 F g^{-1}. Leaf extracts are widely used for the green synthesis of metallic nanoparticles. In this process the reactive organic groups in the extracts are utilized as reducing and stabilizing for metals. *Aloe vera* parts are also used for this purpose. Similar to this approach NiO is modified with the biomolecules in the *Aloe vera* extract to enhance the surface properties. Apart from the above mentioned carbonaceous material synthesis based studies here the composite structure is obtained by the modification of the NiO mas a metal oxide with the biomolecules. Resulted material improved the anodic and cathodic peak potentials of the NiO and provided longer stability and charge–discharge capacity [62]. He et al. [63] reported an interesting study on the effect of the mixture usage as raw biomass material. The raw biomass composed of, rice husk, reed rod, Platanus fruit, fibers, flax fiber, and walnut Shell. The mixture of these materials is rinsed and grinded than calcinated under nitrogen atmosphere finally Hierarchical porous hollow carbon nanospheres (HCNSs) were fabricated. This one step process is also performed by the addition of polytetrafluoroethylene (PTFE) to raw biomass. Both the hollow carbons served well as a supercapacitor additive since they possess core-shell pores. Also silica content improved the mesoporosity of the structure very much.

4. Conclusion

The utilization of the carbon based materials in the improvement of the electrochemical performances of energy production and storage devices has reached an important stage. In this manner, the source depletion for the synthesis of these materials leads the researchers to find new and cost-effective solutions. Today's studies show that biomass provides a real ocean to overcome this problem with many advantages. Especially supercapacitors need reliable modifications in which biomass derived carbon based materials play a crucial role. After extensive investigations biomass is found to be capable of the synthesis of highly porous carbon materials with low/no-cost and eternal precursor supplementary. At this point it has to be underlined that these methods not only provide a way to produce high-value added materials but also contribute to the recycling of the wastes with the win-win principle. The reported studies prove the developed biomass derived materials enhance electrochemical adhesion of the ions which leads to increased specific capacitance of the electrode, consecutively cycling ability and stability of the supercapacitor is enhanced. The crucial points in the biomass derived production are indicated as the choice of the biomass, synthesis process, pretreatment, and the type of the supercapacitor. Among them precursor material and pretreatment play a key role because the pore size and distribution vary very much depending on the precursor content and the pretreatment process. Heteroatom doping to the biomass derived materials add extremely high conductivity to the ordinary materials so the composite materials are preferred in many supercapacitor applications. However, because of the biomass derived synthesis is a green synthesis method the researchers avoid to exaggerated hazardous chemical pretreatments, instead they should choose the right precursor biomass material that has this feature in itself that is reported in this study as well. Overall, biomass is a very valuable source for the synthesis of the next generation of low-cost and green electrode materials for supercapacitors, fuel cells, batteries, and all electrochemical transducers.

Conflict of interest

The authors declare no conflict of interest.

Author details

Sema Aslan[1] and Derya Bal Altuntaş[2*]

1 Department of Chemistry, Faculty of Science, Mugla Sitki Kocman University, Mugla, Turkey

2 Department of Bioengineering, Faculty of Architecture and Engineering, Recep Tayyip Erdogan University, Rize, Turkey

*Address all correspondence to: derya.balaltuntas@erdogan.edu.tr

References

[1] Yang F, Meerman JC, Faaij APC. Carbon capture and biomass in industry: A techno-economic analysis and comparison of negative emission options. Renewable and Sustainable Energy Reviews. 2021;144:111028. https://doi.org/10.1016/j.rser.2021.111028

[2] Antar M, Lyu D, Nazari M, Shah A, Zhou X, Smith DL. Biomass for a sustainable bioeconomy: An overview of world biomass production and utilization. Renewable and Sustainable Energy Reviews. 2021;139:110691. https://doi.org/10.1016/j.rser.2020.110691.

[3] Chen WH, Lin BJ, Lin YY, Chu YS, Ubando AT, Show PL, Ong HC, Chang J-S, Ho S-H, Culaba AB, Pétrissans A, Pétrissans M. Progress in biomass torrefaction: Principles, applications and challenges. Progress in Energy and Combustion Science. 2021;82:100887. https://doi.org/10.1016/j.pecs.2020.100887

[4] Bastida F, Eldridge DJ, García C, Png GK, Bardgett RD, Delgado-Baquerizo M. Soil microbial diversity–biomass relationships are driven by soil carbon content across global biomes. The ISME Journal. 2021;1-11. doi: 10.1038/s41396-021-00906-0.

[5] Di Blasi C. Modeling chemical and physical processes of wood and biomass pyrolysis. Progress in energy and combustion science. 2008;34(1):47-90. https://doi.org/10.1016/j.pecs.2006.12.001

[6] Vassilev SV, Vassileva CG, Vassilev VS. (). Advantages and disadvantages of composition and properties of biomass in comparison with coal: An overview. Fuel. 2015;158:330-350. https://doi.org/10.1016/j.fuel.2015.05.050

[7] Hoa LQ, Vestergaard MDC, Tamiya E. Carbon-based nanomaterials in biomass-based fuel-fed fuel cells. Sensors. 2017;17(11):2587. doi: 10.3390/s17112587.

[8] Tan Z, Yang J, Liang Y, Zheng M, Hu H, Dong H, Liu Y, Xiao Y. The changing structure by component: Biomass-based porous carbon for high-performance supercapacitors. Journal of Colloid and Interface Science. 2021;585:778-786. https://doi.org/10.1016/j.jcis.2020.10.058

[9] Arunachellan IC, Sypu VS, Kera NH, Pillay K, Maity A. Flower-like structures of carbonaceous nanomaterials obtained from biomass for the treatment of copper ion-containing water and their re-use in organic transformations. Journal of Environmental Chemical Engineering. 2021;9(4):105242. https://doi.org/10.1016/j.jece.2021.105242

[10] Gogotsi Y. Nanomaterials handbook. 2nd ed. CRC press; 2017. 712 p. ISBN 9781498703062

[11] Ozin GA, Arsenault A, Cademartiri L. Nanochemistry: a chemical approach to nanomaterials. 2nd ed. Royal Society of Chemistry; 2015. 978-1-84755-895-4

[12] dos Reis GS, de Oliveira H, Larsson SH, Thyrel M, Claudio Lima E. A Short Review on the Electrochemical Performance of Hierarchical and Nitrogen-Doped Activated Biocarbon-Based Electrodes for Supercapacitors. Nanomaterials. 2021;11:424. DOI: 10.3390/nano11020424

[13] Zhang LL, Zhao XS. Carbon-based materials as supercapacitor electrodes. Chemical Society Reviews, 2009;38(9):2520-2531. https://doi.org/10.1039/B813846J

[14] Wang J, Kaskel S. KOH activation of carbon-based materials for energy

storage. Journal of Materials Chemistry. 2012;22(45):23710-23725. https://doi.org/10.1039/C2JM34066F

[15] Wang T, Zang X, Wang X, Gu X, Shao Q, Cao N. Recent advances in fluorine-doped/fluorinated carbon-based materials for supercapacitors. Energy Storage Materials. 2020;30:367-384. https://doi.org/10.1016/j.ensm.2020.04.044

[16] Huang Y, Liang J, Chen Y. An overview of the applications of graphene-based materials in supercapacitors. Small. 2012;8(12):1805-1834. http://dx.doi.org/10.1002/smll.201102635

[17] Zhang LL, Zhou R, Zhao XS. Graphene-based materials as supercapacitor electrodes. Journal of Materials Chemistry, 2010;20(29):5983-5992. https://doi.org/10.1039/C000417K

[18] He Y, Wang L, Jia D, Zhao Z, Qiu J. $NiWO_4$/Ni/Carbon composite fibres for supercapacitors with excellent cycling performance. Electrochimica Acta. 2016;222:446-454. DOI: 10.1016/j.electacta.2016.10.197

[19] Bal Altuntaş D, Aslan S, Akyol Y, Nevruzoğlu V. Synthesis of new carbon material produced from human hair and its evaluation as electrochemical supercapacitor. Energy Sources Part A-Recovery Utilization And Environmental Effects. 2020;42:2346-2356. https://doi.org/10.1080/15567036.2020.1782536

[20] Bal Altuntaş D, Nevruzoğlu V, Dokumacı M, Cam Ş. Synthesis and characterization of activated carbon produced from waste human hair mass using chemical activation. Carbon Letters. 2020;30:307-313. http://dx.doi.org/10.1007/s42823-019-00099-9

[21] Bal Altuntas D, Akgul G, Yanık J, Anık Ü. A biochar-modified carbon paste electrode. Turkish Journal of Chemistry. 2017;41:455-465. doi:10.3906/kim-1610-8

[22] Bal Altuntaş D, Nevruzoğlu V. Evaluation of waste human hair as graphene oxide and examination of some characteristics properties. El-Cezerî Journal of Science and Engineering. 2020;7:104-110. 2020 https://doi.org/10.31202/ecjse.594819

[23] Long, W., Fang, B., Ignaszak, A., Wu, Z., Wang, Y. J., & Wilkinson, D. (2017). Biomass-derived nanostructured carbons and their composites as anode materials for lithium ion batteries. Chemical society reviews, 46(23), 7176-7190. https://doi.org/10.1039/C6CS00639F

[24] Wang, J., Nie, P., Ding, B., Dong, S., Hao, X., Dou, H., & Zhang, X. (2017). Biomass derived carbon for energy storage devices. Journal of materials chemistry a, 5(6), 2411-2428. https://doi.org/10.1039/C6TA08742F

[25] Wang Z, Shen D, Wu C, Gu S. State-of-the-art on the production and application of carbon nanomaterials from biomass. Green Chemistry. 2018;20(22):5031-5057. https://doi.org/10.1039/C8GC01748D

[26] Ellabban O, Abu-Rub H, Blaabjerg F. Renewable energy resources: Current status, future prospects and their enabling technology. Renewable and Sustainable Energy Reviews. 2014;39:748-764. https://doi.org/10.1016/j.rser.2014.07.113

[27] Ibrahim H, Ilinca A, Perron J. Energy storage systems—Characteristics and comparisons. Renewable and sustainable energy reviews. 2008;12(5):1221-1250. https://doi.org/10.1016/j.rser.2007.01.023

[28] Zhang S, Pan, N. Supercapacitors performance evaluation. Advanced Energy Materials. 2015;5(6):1401401. DOI: 10.1002/aenm.201401401

[29] Frackowiak E, Metenier K, Bertagna V, Beguin F. Supercapacitor electrodes from multiwalled carbon nanotubes. Applied Physics Letters. 2000;77(15):2421-2423. https://doi.org/10.1063/1.1290146

[30] Kim J, Lee J, You J, Park MS, Al Hossain MS, Yamauchi Y, Kim JH. Conductive polymers for next-generation energy storage systems: recent progress and new functions. Materials Horizons. 2016;3(6):517-535. https://doi.org/10.1039/C6MH00165C

[31] Aslan S, Bal Altuntaş D, Koçak Ç, Kara Subaşat H. Electrochemical evaluation of Titanium (IV) Oxide/ Polyacrylonitrile electrospun discharged battery coals as supercapacitor electrodes. Electroanalysis. 2021;33:120-128. DOI: 10.1002/elan.202060239

[32] Xu B, Yue S, Sui Z, Zhang X, Hou S, Cao G, Yang Y. What is the choice for supercapacitors: graphene or graphene oxide?. Energy & Environmental Science. 2011;4(8):2826-2830. DOI: 10.1039/c1ee01198g

[33] Singh PK, Das AK, Hatui G, Nayak GC. Shape controlled green synthesis of CuO nanoparticles through ultrasonic assisted electrochemical discharge process and its application for supercapacitor. Materials Chemistry and Physics. 2017;198:16-34. ISSN 0254-0584, https://doi.org/10.1016/j.matchemphys.2017.04.070.

[34] Bose S, Kuila T, Mishra AK, Rajasekar R, Kim NH, Lee JH. (). Carbon-based nanostructured materials and their composites as supercapacitor electrodes. Journal of Materials Chemistry. 2012;22(3):767-784. https://doi.org/10.1039/C1JM14468E

[35] Rivera-Cárcamo C, Serp P. Single atom catalysts on carbon-based materials. ChemCatChem. 2018;10(22):5058-5091. https://doi.org/10.1002/cctc.201801174

[36] Senthil RA, Yang V, Pan J, Sun Y. A green and economical approach to derive biomass porous carbon from freely available feather finger grass flower for advanced symmetric supercapacitors. Journal of Energy Storage. 2021;35:102287. https://doi.org/10.1016/j.est.2021.102287

[37] Zheng LH, Chen MH, Liang SX, Lü QF. Oxygen-rich hierarchical porous carbon derived from biomass waste-kapok flower for supercapacitor electrode. Diamond and Related Materials. 2021;113:108267. https://doi.org/10.1016/j.diamond.2021.108267

[38] Jain A, Ghosh M, Krajewski M, Kurungot S, Michalska M. Biomass-derived activated carbon material from native European deciduous trees as an inexpensive and sustainable energy material for supercapacitor application. Journal of Energy Storage. 2021;34:102178. https://doi.org/10.1016/j.est.2020.102178

[39] Nguyen NT, Le PA, Phung VBT. Biomass-derived carbon hooks on Ni foam with free binder for high performance supercapacitor electrode. Chemical Engineering Science. 2021;229:116053. https://doi.org/10.1016/j.ces.2020.116053

[40] Long S, Feng Y, He F, Zhao J, Bai T, Lin H, Cai W, Mao C, Chen Y, Gan L, Liu J, Ye M, Zeng X, Long M. Biomass-derived, multifunctional and wave-layered carbon aerogels toward wearable pressure sensors, supercapacitors and triboelectric nanogenerators. Nano Energy. 2021;85:105973. https://doi.org/10.1016/j.nanoen.2021.105973

[41] Wu Y, Cao J-P, Zhuang Q-Q, Zhao X-Y, Zhou Z, Wei Y-L, Zhao M, Bai H-C. Biomass-derived three-dimensional hierarchical porous carbon network for symmetric supercapacitors with ultra-high energy density in ionic liquid electrolyte, Electrochimica Acta.

2021;371:137825. https://doi.org/10.1016/j.electacta.2021.137825.

[42] Fang C, Hu P, Dong S, Cheng Y, Zhang D, Zhang X. Construction of carbon nanorods supported hydrothermal carbon and carbon fiber from waste biomass straw for high strength supercapacitor. Journal of Colloid and Interface Science. 2021;582:552-560. https://doi.org/10.1016/j.jcis.2020.07.139

[43] Ba H, Wang W, Pronkin S, Romero T, Baaziz W, Nguyen-Dinh L, Chu W, Ersen O, Huu CP. Biosourced foam-like activated carbon materials as high-performance supercapacitors. Adv Sustain Syst. 2018;2:1700123. https://doi.org/10.1002/adsu.201700123

[44] Wang B, Ji L, Yu Y, Wang N, Wang J, Zhao J. A simple and universal method for preparing N, S co-doped biomass derived carbon with superior performance in supercapacitors. Electrochim Acta. 2019;309:34-43. https://doi.org/10.1016/j.electacta.2019.04.087

[45] Liu Y, Yu W, Hou L, He G, Zhu Z. Co_3O_4@Highly ordered macroporous carbon derived from a mollusc shell for supercapacitors. RSC Adv 2015;5:75105-75110. DOI: 10.1039/C5RA15024H

[46] Xiong W, Gao Y, Wu X, Hu X, Lan D, Chen Y, Pu X, Zeng Y, Su J, Zhu Z. Composite of macroporous carbon with honeycomb-like structure from mollusc shell and NiCo$_{(2)}$O$_{(4)}$ nanowires for high-performance supercapacitor. ACS Appl Mater Interfaces. 2014;6:19416-19423. https://doi.org/10.1021/am5055228

[47] Lai F, Miao YE, Zuo L, Lu H, Huang Y, Liu T. Biomass-derived nitrogen-doped carbon nanofiber network: a facile template for decoration of ultrathin nickelcobalt layered double hydroxide nanosheets as high-performance asymmetric supercapacitor electrode. Small. 2016;12:3235-3244. https://doi.org/10.1002/smll.201600412

[48] Ghosh S, Santhosh R, Jeniffer S, Raghavan V, Jacob G, Nanaji K, Kollu P, Jeong SK, Grace AN. Natural biomass derived hard carbon and activated carbons as electrochemical supercapacitor electrodes. Sci Rep. 2019;9:1-15. https://doi.org/10.1038/s41598-019-52006-x.

[49] Shang T, Xu Y, Li P, Han J, Wu Z, Tao Y, Yang QH. A bioderived sheet-like porous carbon with thin-layer pore walls for ultrahigh-power supercapacitors. Nanomater Energy. 2020;70:104531. https://doi.org/10.1016/j.nanoen.2020.104531.

[50] He J, Zhang D, Wang Y, Zhang J, Yang B, Shi H, Wang K, Wang Y. Biomass-derived porous carbons with tailored graphitization degree and pore size distribution for supercapacitors with ultra-high rate capability. Appl Surf Sci. 2020;515:146020. https://doi.org/10.1016/j.apsusc.2020.146020.

[51] Selvaraj AR, Muthusamy A, Kim HJ, Senthil K, Prabakar K. Ultrahigh surface area biomass derived 3D hierarchical porous carbon nanosheet electrodes for high energy density supercapacitors. Carbon. 2021;174:463-474. https://doi.org/10.1016/j.carbon.2020.12.052

[52] Cao X, Li Z, Chen H, Zhang C, Zhang Y, Gu C, Xu X, Li Q. Synthesis of biomass porous carbon materials from bean sprouts for hydrogen evolution reaction electrocatalysis and supercapacitor electrode. International Journal of Hydrogen Energy. https://doi.org/10.1016/j.ijhydene.2021.03.038

[53] Yakaboylu GA, Jiang C, Yumak T, Zondlo JW, Wang J, Sabolsky EM. Engineered hierarchical porous carbons for supercapacitor applications through chemical pretreatment and activation of biomass precursors. Renewable

Energy.2021;163:276-287. https://doi.org/10.1016/j.renene.2020.08.092

[54] Chaparro-Garnica J, Salinas-Torres D, Mostazo-López MJ, Morallón E, Cazorla-Amorós D. Biomass waste conversion into low-cost carbon-based materials for supercapacitors: A sustainable approach for the energy scenario. Journal of Electroanalytical Chemistry. 2021;880:114899. http://dx.doi.org/10.1016/j.jelechem.2020.114899

[55] Liang X, Liu R, Wu X. Biomass waste derived functionalized hierarchical porous carbon with high gravimetric and volumetric capacitances for supercapacitors. Microporous and Mesoporous Materials. 2021;310:110659. https://doi.org/10.1016/j.micromeso.2020.110659

[56] Du J, Zhang Y, Lv H, Chen A. Silicate-assisted activation of biomass towards N-doped porous carbon sheets for supercapacitors. Journal of Alloys and Compounds. 2021;853:157091. https://doi.org/10.1016/j.jallcom.2020.157091

[57] Ariharan A, Ramesh K, Vinayagamoorthi R, Rani MS, Viswanathan B, Ramaprabhu S, Nandhakumar V. Biomass derived phosphorous containing porous carbon material for hydrogen storage and high-performance supercapacitor applications. Journal of Energy Storage. 2021;35:102185. https://doi.org/10.1016/j.est.2020.102185

[58] Ramanathan S, Sasikumar M, Paul SPM, Obadiah A, Angamuthu A, Santhoshkumar P, Raj D, Vasanthkumar S. Low cost electrochemical composite material of paper cup waste carbon (P-carbon) and Fluorescein for supercapacitor application. Materials Today: Proceedings. https://doi.org/10.1016/j.matpr.2020.12.561

[59] Xie X, Zhang B, Wang Q, Zhao X, Wu D, Wu H, Sun X, Hou C, Yang X, Yu R, Zhang S, Murugadoss V, Du W. Efficient microwave absorber and supercapacitors derived from puffed-rice-based biomass carbon: Effects of activating temperature. Journal of Colloid and Interface Science. 2021;594:290-303. https://doi.org/10.1016/j.jcis.2021.03.025

[60] Ajay KM, Dinesh MN, Byatarayappa G, Radhika MG, Kathyayini N, Vijeth H. Electrochemical investigations on low cost KOH activated carbon derived from orange-peel and polyaniline for hybrid supercapacitors. Inorganic Chemistry Communications. 2021;127:108523. https://doi.org/10.1016/j.inoche.2021.108523

[61] Gehrke V, Maron GK, da Silva Rodrigues L, Alano JH, de Pereira CMP, Orlandi MO, Carreño NLV. Facile preparation of a novel biomass-derived H3PO4 and Mn $(NO_3)_2$ activated carbon from citrus bergamia peels for high-performance supercapacitors. Materials Today Communications. 2021;26:101779. https://doi.org/10.1016/j.mtcomm.2020.101779

[62] Avinash B, Ravikumar CR, Kumar MA, Santosh MS, Pratapkumar C, Nagaswarupa HP, Ananda Murthy HC, Deshmukh VV, Bhatt Aarti S, Jahagirdar AA, Alam MW. NiO bio-composite materials: Photocatalytic, electrochemical and supercapacitor applications. Applied Surface Science Advances. 2021;3:100049. https://doi.org/10.1016/j.apsadv.2020.100049

[63] He D, Gao Y, Wang Z, Yao Y, Wu L, Zhang J, Huang Z-H, Wang MX. One-step green fabrication of hierarchically porous hollow carbon nanospheres (HCNSs) from raw biomass: Formation mechanisms and supercapacitor applications. Journal of Colloid and Interface Science. 2021;581:238-250. https://doi.org/10.1016/j.jcis.2020.07.118

Section 3

Graphene Based Materials

Chapter 4

Graphene-Based Materials for Supercapacitor

Abu Jahid Akhtar

Abstract

Graphene, a one-atomic-thick film of two-dimensional nanostructure, has piqued the attention of researchers due to its superior electrical conductivity, large surface area, good chemical stability, and excellent mechanical behaviour. These extraordinary properties make graphene an appropriate contender for energy storage applications. However, the agglomeration and re-stacking of graphene layers due to the enormous interlayer van der Waals attractions have severely hampered the performance of supercapacitors. Several strategies have been introduced to overcome the limitations and established graphene as an ideal candidate for supercapacitor. The combination of conducting polymer (CP) or metal oxide (MO) with graphene as electrode material is expected to boost the performance of supercapacitors. Recent reports on various CP/graphene composites and MO/graphene composites as supercapacitor electrode materials are summarised in this chapter, with a focus on the two basic supercapacitor mechanisms (EDLCs and pseudocapacitors).

Keywords: Supercapacitor, EDLC, Pseudo-capacitance, Metal oxides, Conducting polymers

1. Introduction

Energy storage devices are important in today's world to meet the increasing demand for reliable and portable power sources [1–3]. Supercapacitors, also known as ultracapacitors, are electrochemical energy storage devices that are lightweight, can operate at a wide range of temperatures, have a long life cycle, and are shielded to make their work easier [4, 5]. With a number of such advantages, the supercapacitors emerged in a variety of applications in hybrid or electric vehicles, electronics and aircrafts [4, 5]. Today, supercapacitor manufacturers mostly use coconut activated carbon as an electrode material due to its high specific surface area, low price and mass production capability. However, with increased energy demand, significant research efforts have been made to find ideal electrode materials for the production of advanced energy storage systems. Energy is stored in supercapacitors via two energy storage mechanism, namely electrochemical double layer capacitance (EDLC) and pseudocapacitance. So, in order to improve supercapacitor efficiency, both of these mechanisms must be incorporated on a single electrode material.

Graphene, a one-atom-thick 2D single layer of sp^2 -bonded carbon atoms with the hexagonal lattice structure, is considered as the basic building block material for all carbon materials [6, 7]. Graphene has emerged as an appropriate candidate for energy storage applications due to its high electrical (10^8 S/m) and thermal

IntechOpen

conductivity (5000 W/m/K), large surface area ($2.63X10^6 m^2/kg$), high transparency (absorbance of 2.3%), good chemical stability, and excellent mechanical behaviour (breaking strength of 42 N/m and Young's modulus of 1.0 TPa) [8, 9]. Various methods such as chemical vapour deposition (CVD) of hydrocarbons, epitaxial growth on electrically insulating surfaces such as SiC, micromechanical exfoliation of graphite (Scotch tape method), oxidation–exfoliation–reduction of graphite powder may be used to synthesise graphene sheets of various sizes and defect contents [10]. Among these methods, graphene sheet which is grown by chemical vapour deposition (CVD) of hydrocarbon [11] has the superior quality with minimal defects. However, CVD prepared graphene would not be an ideal contender for EDLC electrode material, as it is too costly to produce and is hardly scalable. On the other hand, graphene produced by a chemical or thermal exfoliation process [12] of graphite is relatively inexpensive but has more surface defects which prevent graphene from being used in high-speed electronic, photonic/optoelectronic devices. But these defects play an important role in supercapacitor applications. So graphene with defects is used as an acceptable supercapacitor material. However, due to the large interlayer van der Waals attractions, re-stacking of graphene layers can severely reduce the available electrochemical surfaces, obstructing ion diffusion and ultimately limiting electrochemical efficiency, and the lack of fast Faradic pseudocapacitive behaviour have severely hampered supercapacitor performance. In order to address the issue graphene is often combined with other materials such as different metal oxides (MP) and conducting polymers (CP) to further increase its electrochemical performance. Coupling MPs or CPs to graphene has been shown to be an effective approach to improving the cycling stability, energy and power density of the supercapacitor device by introducing pseudocapacitance [13–37]. This chapter summarises recent studies on various CP/graphene composites and MO/graphene composites as supercapacitor electrode materials, with an emphasis on the two basic supercapacitor mechanisms (EDLCs and pseudocapacitors).

2. Graphene-metal oxide nanocomposites for supercapacitor applications

Metal oxide supercapacitors have gotten a lot of attention in recent years because of their high theoretical basic capacitance, low cost, environmental friendliness, and natural abundance [38–45]. Metal oxides also allow rapid, reversible faradic reactions to the electrode-electrolyte interface [46] resulting in large specific capacitances. However, the power density and cycling stability of the metal oxide based supercapacitor device are limited by poor electronic and ionic conductivity of metal oxides. So metal oxides are often combined with graphene to address these drawbacks, and it is expected that hybrid metal oxide/graphene nanostructures can increase supercapacitor performance for large-scale energy storage systems.

2.1 Graphene-manganese oxide nanocomposites

Manganese oxide-graphene composite is the most studied electrode material for supercapacitor devices among metal oxides [13–15]. The charge storage mechanism of a MnO2 electrode involves a transition in manganese (Mn) oxidation state from III to IV. The reversible insertion/extraction of electrolyte cations to balance the charge during reduction/oxidation of Mn^{+3}/Mn^{+4} gives MnO2 its pseudocapacitive properties [47, 48].

He et al. [13] used electrodeposition to build freestanding, lightweight (0.75 mg/cm^2), ultrathin (<200 μm), highly conductive (55 S/cm), and flexible three-dimensional (3D) graphene networks filled with MnO_2 as the flexible supercapacitor electrode material. The composite with 9.8 mg/cm^2 MnO_2mass loading (92.9% of the total electrode mass) had a capacitance of 1.42 F/cm^2 in a scan rate of 2 mV/s. He et al. further optimised the MnO_2 content in the composite material for realistic applications and achieved a maximum specific capacitance of 130 F/g.

Another research [14] rendered graphene/ MnO_2 composites by chemically reducing GO/ MnO_2 with both hydrazine hydrate (H-RGO/MnO_2) and sodium borohydride (S-RGO/ MnO_2) as reducing agents. The H-RGO/MnO_2 showed a specific capacitance of 327.5 F/g, which is higher than that of the S-RGO/MnO_2 (278.6 F/g). Kim et al. proposed that using the hydrazine reduction process to fabricate MnO_2 on graphene oxide surfaces is a promising fabrication method for supercapacitor electrodes.

For producing highly efficient graphene/metal oxide-based hybrid supercapacitors, Wang et al. [15] described in situ synthesis of 3D-graphene/MnO2 foam composite using a combination of chemical vapour deposition and hydrothermal process. High crystallinity and low contact resistance were observed during in situ conformal growth of 3D-graphene/MnO_2 composites. In the supercapacitor, the 3D-graphene/MnO_2 composite electrode demonstrated high specific capacitance (333.4 F/g at 0.2 A/g) and excellent cycling stability (92.2% retention at 0.2 A/g after 2000 cycles).

Thus these methods for fabricating graphene/MnO_2 composites offers a promising means of producing energy storage electrode materials for supercapacitor devices with high efficiency.

2.2 Graphene-iron oxide nanocomposites

Iron oxides drew interest as a potential electrode material due to their natural abundance, high thermal stability, and low toxicity [42, 43]. However, in terms of power density and cyclic stability, iron oxide struggles as an electrode material and must be combined with graphene to overcome the problem.

Reduced graphene oxide- Fe_3O_4 (RGO-Fe_3O_4) nanocomposite was synthesised by Ghasemi and co-worker [16] using a simple electrophoretic deposition (EPD) method followed by an electrochemical reduction procedure. On RGO, Fe_3O_4 nanoparticles with a diameter of 20–50 nm are uniformly assembled. At a current density of 1 A/g, RGO- Fe_3O_4 had a specific capacitance of 154 F/g, which is greater than RGO (81 F/g) in Na_2SO_4 electrolyte. The electrochemical behaviours of this study also revealed that adding surfactant to aqueous Na_2SO_4 solution would boost the capacitance of RGO- Fe_3O_4 electrodes. RGO- Fe_3O_4 electrode in Na2SO4 electrolyte containing t-octyl phenoxy polyethoxyethanol (Triton X-100) showed capacitance of 236 F/g at 1 A/g, with 97% of the initial capacitance retained after 500 cycles.

Qu et al. [17] have shown that the increase in electrochemical capacitive performance of 2D Fe_3O_4 -graphene nanocomposites was mainly due to the optimization of electrochemical surfaces by avoiding graphene re-stacking due to the uniform Fe_3O_4 surface deposition and synergistic effect of Fe_3O_4 and graphene. The hybrid capacitor shows a capacitance value of 304 F/g. In addition, Fe_3O_4 -graphene nanocomposites have achieved higher power density.

In another work Tang et al. [18] synthesised three-dimensional (3D) iron oxide/graphene aerogel hybrid using an innovative in situ hydrothermal process for supercapacitor applications. This material Fe_2O_3/GA hybrid electrode were used to make a highly flexible all-solid-state symmetric supercapacitor system.

The device provided a high specific capacity of 440 F/g and was suitable for different bending angles. 90% of the capacitance was also preserved after 2200 cycles, indicating strong cycling stability. These excellent electrochemical performances suggest that graphene-iron oxide nanocomposites have huge potential in energy application.

2.3 Graphene-nickel oxide nanocomposites

Nickel oxide (NiO) has been shown to be one of the most promising electrode materials for supercapacitors [38, 39]. However, the efficacy of NiO has been found to be reduced due to its low electrical conductivity, resulting in poor performance in electrochemical devices. Researchers are trying to boost its efficiency by linking it to graphene.

A simple solvothermal-induced self-assembly method was used by Gui and co-worker [19] to make three-dimensional nickel oxide/graphene aerogel nanocomposites (NiO/GA). With an extremely large working potential window, the NiO/GA electrodes attained a specific capacitance of 587.3 F/g at 1A/g. The NiO/GA had excellent cycling reliability, with only a minor decrease in capacitance after 1000 cycles.

Zhao et al. [20] demonstrated NiO's electrochemical properties by growing NiO mesoporous nanowalls on rGO nanosheets on 3D nickel foams, referring to the process as binder-free electrode preparation. The NiO-graphene nanocomposite 3D porous foam composite substrate offers an appropriate structure for electron collection and electrolyte/ion diffusion via the active materials, resulting in a high specific capacitance of 950 F/g at a current density of 5 A/g with excellent cycling stability.

In another work Choi et al. [21] demonstrated the synthesis of 3D porous graphene/NiO nanoparticle composites (3D-RGNi) by a facile method. The prepared 3D-RGNi had a large electrochemically active surface region. The as-synthesised 3D-RGNi electrode had a large specific capacitance of 1328 F/g at 1A/g and superb rate capability, with 87% of the capacitance retained after 2000 cycles. The synergistic effects of the rGO network and NiO nanoparticles, as well as the highly porous structure of 3D-RGNi, are attributed with these high capacitance results.

Hence, it is expected that graphene-Nickel oxide nanocomposites might serve as a favourable materials for energy storage applications.

2.4 Graphene-cobalt oxide nanocomposites

Apart from having a high theoretical capacitance (3560 F/g) and being abundant in nature, cobalt-oxide based nanomaterials are also said to be environmentally friendly [49, 50]. The faradaic redox transitions of interfacial oxy-cation species trigger the pseudo capacitance of hydrous Cobalt oxides [51]. The formation of cobalt oxide phases with a transition between Co(II), Co(III), and Co(IV) oxidation states explains the charge–discharge process of cobalt oxide in alkaline electrolyte.

Akhtar et al. [22] presented the preparation of nanostructured cobalt oxide/reduced graphene oxide (Co_3O_4/rGO) composites for potential materials in supercapacitor applications using a basic one-step cost-effective hydrothermal technique. The v nanoparticles in the Co_3O_4/rGO nanocomposites were layered over the surface of the rGO sheets. In a three electrode cell system, Co_3O_4/rGO nanocomposite based electrode had a specific capacitance of 754 F/g and after 1000 continuous cycles, the material abled to maintained 96% of its initial capability.

Chen et al. [23] used a simple hydrothermal technique to manufacture cobalt oxide (Co_3O_4) nanowires on three-dimensional (3D) graphene foam. The free-standing electrode for supercapacitor application was prepared from the synthesised 3D graphene/Co_3O_4 composite. It showed a high specific capacitance of 1100 F/g and excellent cycling stability at a current density of 10 A/g.

3. Graphene-conducting polymers for supercapacitor applications

Conducting polymers (CPs) attains lots of attention in academia and industry as electrode materials for supercapacitor. Polyaniline (PANI), polypyrrole (PPy) and poly(3,4-ethylenedioxythiophene) polystyrene sulfonate (PEDOT:PSS) are type of conducting polymers [34, 52–56] that are extensively studied for supercapacitor device by offering very fast redox reaction with an electrolyte which can lead to pseudo-capacitance. However, CPs have a disadvantage of long-term stability due to the mechanical degradation of CPs. Thus graphene is often used to overcome the issue with CPs and guide to long term cycling stability, which is vital for supercapacitor devices.

3.1 Graphene- polyaniline (PANI) composite

PANI exhibit excellent conductivity and stability and have been widely used in energy storage devices [52, 53]. During the last few years, graphene/PANI composite nanocomposites have been used as electrode materials for supercapacitors [24–28]. Wu et al. [24] demonstrated synthesis of polyaniline nanofiber on graphene by in situ polymerisation of aniline monomer in the presence of graphene oxide under acid condition. The supercapacitor devices shows a specific capacitance of 480 F/g at a current density of 0.1 A/g and the material retained 70% of the original capacitance value after 1000 cycles. Gomes et al. [25] aimed to improve the cyclic stability by preparing hierarchical assembly of graphene/polyaniline nanostructures by microemulsion polymerisation, followed by the incorporation of graphene oxide nanosheets by hierarchical organisation. Hierarchical nanostructures showed a specific capacitance of 448F/g which is almost double of that of PANI due to the synergistic combination of graphene and PANI nanostructures. At the same time almost 81% capacity retention was achieved for the material compared to 38% for PANI after 5000 cyclic operations. Cheng et al. [26] prepared graphene–PANI composite paper as a flexible electrode, by combining the advantages of high conductivity, mechanical strength, and flexibility of graphene paper and large capacitance of the PANI. Based on these properties, this flexible graphene–PANI electrode material displayed a good tensile strength of 12.6 MPa and a stable large electrochemical capacitance of 233 F/g and 135 F/cm3 for gravimetric and volumetric capacitances, respectively.

Zhang et al. [27] reported a novel method to prepare flexible graphene/polyaniline paper (GPp) as supercapacitor electrodes through controlled in-situ polymerisation followed by roll coating in order to increase the electrochemical properties. This GPp deliver a high specific capacitance of 838 F/g at a current density of 1 A/g and high retention of 93.7% at 10 A/g over 5000 cycles. Kinetics analysis of the material shows that the GPp stores both surface capacitance and diffusion capacitance. The as-prepared GPp also showed a specific energy density as high as 40 Wh/kg and a power density of 10 kW/kg. The authors also succeeded to light the light emitting diode (LED) connected with the fabricated GPp device.

Another method which is commonly used for the preparation of graphene polymer composite is physical mixing in a given solvent. Flexible Graphene/Polyaniline Nanofiber Composite films were prepared by vacuum filtration of the mixed dispersions of both rGO and PANI nanofibers [28]. The film was mechanically stable and showed a good flexibility. The supercapacitor devices showed large electrochemical capacitance of 210 F/g at a low discharge rate of 0.3 A/g.

3.2 Graphene-Polypyrrole (PPy)

PPy is emerged as attracting material for supercapacitor [54, 55] because of its simple synthesis procedure and it has good thermal and electrical conductivity. However, the thin PPy film forms aggregated a cauliflower-like structure which is not favourable for supercapacitor applications. In order to address the issue PPy is often combined with graphene to improve its electrochemical performance. Zhang et al. [29] synthesised graphene and polypyrrole composite via in situ polymerisation of pyrrole monomer in the presence of graphene under acid conditions An even composite is formed with polypyrrole being uniformly bounded by graphene sheets. Electrochemical performance of the composite material are higher than pure samples with the maximum capacitance of 482 F/g and excellent cycling performance (95% retention after 1000 cycles). Liu et al. [30] demonstrated preparation of hierarchical graphene/polypyrrole nanocomposites via in-situ polymerisation of self-assembled pyrrole. The nanomaterial attained outstanding conductivity of ~1980 S/cm and demonstrated promising potential in supercapacitor, with a specific capacitance value 650 F/g at 0.45 A/g current density. Furthermore, the device showed energy density of 54.0 W h/kg at 1 mA current, and power density of 778.1 W/kg at 5 mA current. In another work Akhtar et al. [31] studied the electrochemical performance of graphene/polypyrrole layered type structure. Charge transport was investigated in this study to determine the relative contributions of graphene and polypyrrole in charge transport and storage mechanism, with the aim of improving device properties. Electrochemical supercapacitor fabricated using this layered composite exhibited a large value (~931 F/g) of specific capacitance.

3.3 Graphene/(PEDOT:PSS) composite

PEDOT: PSS attract as one of the potential electrode materials due to its good electrical conductivity, transparency, ductility, and stability [34, 56]. Wu et al. [32] demonstrated ultrathin printable graphene supercapacitors based on solution-processed electrochemically exfoliated graphene hybrid films on an ultrathin poly(ethylene terephthalate) substrate. The device exhibited an unprecedented volumetric capacitance of 348 F/cm3 at an ultrahigh scan rate of 2000 V/s, and AC line-filtering performance. This method can be possibly used for large-scale production of printable, thin and lightweight supercapacitor devices.

Fibre-shaped supercapacitors [33] with high mechanical and electronic properties based on hollow rGO/PEDOT: PSS (HCF) have gained tremendous attention because of their tiny volume, low weight, high flexibility, and good wearability. This novel fibre-shaped supercapacitor showed a high specific capacitance of 304.5 mF/cm2 at 0.08 mA/cm2 and an energy density of 27.1 mWh/cm2. In another work highly flexible, bendable and conductive rGO-PEDOT/PSS films were prepared by Chen et al. [34] The assembled device could be bent and twisted without harming the electrochemical performance of the device. A high areal capacitance of 448 mF/cm2 was achieved at a scan rate of 10 mV/s and when the device was fully charged the device was powerful enough to power a LED for 20 seconds.

3.4 Graphene/other polymers composite

Besides the above-mentioned conducting polymers there are others polymers that have been combined with graphene for supercapacitor devices. Gupta and co-workers [35] synthesised Poly (3-hexylthiophene)/graphene composites via both in-situ and ex-situ growth technique to investigate supercapacitive behaviour. They observed that in-situ growth of P3HT forms better composites with graphene than ex-situ growth. The values of specific capacitance for ex-situ and in-situ samples were found to be 244 F/g and 323 F/g respectively at a current density of 200 mA/g. Thus in-situ P3HT/graphene composite showed superior storage capacity in comparision to ex-situ sample. Electrochemical performance was also studied for graphene (G)–polyethylenedioxythiophene (PEDOT) nanocomposites as electrode material [36]. This manuscript presented the capacitance studies on supercapacitor G-PEDOT electrode with respect to stability of material, specific capacitance and electrical conductivity. Specific capacitance value for G-PEDOT sample was estimated to be 374 F/g. Wu et al. [37] reported conjugated polyfluorene imidazolium ionic liquids (coPIL) intercalated reduced graphene oxide (coPIL-RGO) for high performance supercapacitor. coPIL-RGO based device showed a specific capacitance of 222 F/g at a low current density of 0.2 A/g in 6 M KOH and 132 F/g at a current density of 0.5 A/g in ionic liquid electrolyte 1-butyl-3-methylimidazolium tetrafluoroborate (BMIMBF4), respectively.

4. Conclusion

Metal oxide and Conducting polymers have gotten a lot of attention in next-generation supercapacitor electrode research because of their simple synthesis procedure, low cost and high pseudocapacitance. Simultaneously, pure metal oxide and conducting polymers have a number of flaws, including low electrical conductivity, weak cyclic stability, and low energy and power density. Graphene/conducting polymer composites and graphene/metal oxide composites outperform conducting polymers and metal oxides in terms of cyclic durability, energy density, and power density. Simultaneously, in recent years, considerable focus has been placed on structural architecture, material fabrication, and device performance evaluation. To accomplish the expected full-scale realistic application, both the efficiency and reproducible quantity of the electrode materials must be improved in the immediate future.

Author details

Abu Jahid Akhtar
Department of Physics, Diamond Harbour Women's University, West Bengal, India

*Address all correspondence to: jahid.dhwu@gmail.com

References

[1] Dell RM, Rand DAJ. Energy storage—a key technology for global energy sustainability. Journal of power sources. 2001; 100: 2-17.

[2] Suberu MY, Mustafa MW, Bashir N. Energy storage systems for renewable energy power sector integration and mitigation of intermittency. Renewable Sustainable Energy Reviews. 2014; 35: 499-514.

[3] Yoo HD, Markevich E, Salitra G, Sharon D, Aurbach D. On the challenge of developing advanced technologies for electrochemical energy storage and conversion. Materials Today. 2014; 17: 110-121.

[4] Raza W, Ali F, Raza N, Luo Y, Kim K-H, Yang J, Kumar S, Mehmood A, Kwon EE. Recent advancements in supercapacitor technology. Nano Energy. 2018; 52: 441-473.

[5] Zuo W, Li R, Zhou C, Li Y, Xia J, Liu J. Battery-supercapacitor hybrid devices: recent progress and future prospects. Advanced science. 2017; 4: 1600539.

[6] Dong Y, Wu Z-S, Ren W, Cheng H-M, Bao X. Graphene: a promising 2D material for electrochemical energy storage. Science Bulletin. 2017; 62: 724-740.

[7] Mas-Balleste R, Gomez-Navarro C, Gomez-Herrero J, Zamora F. 2D materials: to graphene and beyond. Nanoscale. 2011; 3: 20-30.

[8] Novoselov KS, Geim AK, Morozov SV, Jiang D, Zhang Y, Dubonos SV, Grigorieva IV, Firsov AA. Electric field effect in atomically thin carbon films. science. 2004; 306: 666-669.

[9] Geim AK, Novoselov KS. The rise of graphene. Nature Materials 2007; 6: 183-191.

[10] Choi W, Lahiri I, Seelaboyina R, Kang YS. Synthesis of graphene and its applications: a review. Critical Reviews in Solid State Materials Sciences. 2010; 35: 52-71.

[11] Li X, Cai W, Colombo L, Ruoff RS. Evolution of graphene growth on Ni and Cu by carbon isotope labeling. Nano letters. 2009; 9: 4268-4272.

[12] Yu H, Zhang B, Bulin C, Li R, Xing R. High-efficient synthesis of graphene oxide based on improved hummers method. Scientific reports. 2016; 6: 1-7.

[13] He Y, Chen W, Li X, Zhang Z, Fu J, Zhao C, Xie E. Freestanding three-dimensional graphene/MnO2 composite networks as ultralight and flexible supercapacitor electrodes. ACS nano. 2013; 7: 174-182.

[14] Kim M, Hwang Y, Kim J. Graphene/MnO2-based composites reduced via different chemical agents for supercapacitors. Journal of power sources. 2013; 239: 225-233.

[15] Bai X-L, Gao Y-L, Gao Z-Y, Ma J-Y, Tong X-L, Sun H-B, Wang JA. Supercapacitor performance of 3D-graphene/MnO2 foam synthesized via the combination of chemical vapor deposition with hydrothermal method. Applied Physics Letters. 2020; 117: 183901.

[16] Ghasemi S, Ahmadi F. Effect of surfactant on the electrochemical performance of graphene/iron oxide electrode for supercapacitor. Journal of Power Sources. 2015; 289: 129-137.

[17] Qu Q, Yang S, Feng X. 2d sandwich-like sheets of iron oxide grown on graphene as high energy anode material for supercapacitors. Advanced materials. 2011; 23: 5574-5580.

[18] Khattak AM, Yin H, Ghazi ZA, Liang B, Iqbal A, Khan NA, Gao Y, Li L, Tang Z. Three dimensional iron oxide/graphene aerogel hybrids as all-solid-state flexible supercapacitor electrodes. RSC advances. 2016; 6: 58994-59000.

[19] Chen W, Gui D, Liu J. Nickel oxide/graphene aerogel nanocomposite as a supercapacitor electrode material with extremely wide working potential window. Electrochimica Acta. 2016; 222: 1424-1429.

[20] Zhao B, Wang T, Jiang L, Zhang K, Yuen MM, Xu J-B, Fu X-Z, Sun R, Wong C-P. NiO mesoporous nanowalls grown on RGO coated nickel foam as high performance electrodes for supercapacitors and biosensors. Electrochimica Acta. 2016; 192: 205-215.

[21] Trung NB, Van Tam T, Dang DK, Babu KF, Kim EJ, Kim J, Choi WM. Facile synthesis of three-dimensional graphene/nickel oxide nanoparticles composites for high performance supercapacitor electrodes. Chemical Engineering Journal. 2015; 264: 603-609.

[22] Sagadevan S, Marlinda AR, Johan MR, Umar A, Fouad H, Alothman OY, Khaled U, Akhtar M, Shahid M. Reduced graphene/nanostructured cobalt oxide nanocomposite for enhanced electrochemical performance of supercapacitor applications. Journal of colloid interface science. 2020; 558: 68-77.

[23] Dong X-C, Xu H, Wang X-W, Huang Y-X, Chan-Park MB, Zhang H, Wang L-H, Huang W, Chen P. 3D graphene–cobalt oxide electrode for high-performance supercapacitor and enzymeless glucose detection. ACS nano. 2012; 6: 3206-3213.

[24] Zhang K, Zhang LL, Zhao X, Wu J. Graphene/polyaniline nanofiber composites as supercapacitor electrodes. Chemistry of Materials. 2010; 22: 1392-1401.

[25] Hassan M, Reddy KR, Haque E, Faisal SN, Ghasemi S, Minett AI, Gomes VG. Hierarchical assembly of graphene/polyaniline nanostructures to synthesize free-standing supercapacitor electrode. Composites science technology. 2014; 98: 1-8.

[26] Wang D-W, Li F, Zhao J, Ren W, Chen Z-G, Tan J, Wu Z-S, Gentle I, Lu GQ, Cheng H-M. Fabrication of graphene/polyaniline composite paper via in situ anodic electropolymerization for high-performance flexible electrode. ACS nano. 2009; 3: 1745-1752.

[27] Yu H, Ge X, Bulin C, Xing R, Li R, Xin G, Zhang B. Facile fabrication and energy storage analysis of graphene/PANI paper electrodes for super capacitor application. Electro chimica Acta. 2017; 253: 239-247.

[28] Wu Q, Xu Y, Yao Z, Liu A, Shi G. Supercapacitors based on flexible graphene/polyaniline nanofiber composite films. ACS nano. 2010; 4: 1963-1970.

[29] Zhang D, Zhang X, Chen Y, Yu P, Wang C, Ma Y. Enhanced capacitance and rate capability of graphene/polypyrrole composite as electrode material for supercapacitors. Journal of Power Sources. 2011; 196: 5990-5996.

[30] Liu Y, Wang H, Zhou J, Bian L, Zhu E, Hai J, Tang J, Tang W. Graphene/polypyrrole intercalating nano composites as supercapacitors electrode. Electrochimica Acta. 2013; 112: 44-52.

[31] Akhtar AJ, Mishra S, Saha SK. Charge transport mechanism in reduced graphene oxide/polypyrrole based ultrahigh energy density supercapacitor. Journal of Materials Science: Materials in Electronics. 2020; 31: 11637-11645.

[32] Wu ZS, Liu Z, Parvez K, Feng X, Müllen K. Ultrathin printable graphene supercapacitors with AC line-filtering performance. Advanced Materials. 2015; 27: 3669-3675.

[33] Qu G, Cheng J, Li X, Yuan D, Chen P, Chen X, Wang B, Peng H. A fiber supercapacitor with high energy density based on hollow graphene/ conducting polymer fiber electrode. Advanced materials. 2016; 28: 3646-3652.

[34] Liu Y, Weng B, Razal JM, Xu Q, Zhao C, Hou Y, Seyedin S, Jalili R, Wallace GG, Chen J. High-performance flexible all-solid-state supercapacitor from large free-standing graphene-PEDOT/PSS films. Scientific reports. 2015; 5: 1-11.

[35] Gupta A, Akhtar AJ, Saha SK. In-situ growth of P3HT/graphene composites for supercapacitor application. Materials Chemistry Physics. 2013; 140: 616-621.

[36] Alvi F, Ram MK, Basnayaka PA, Stefanakos E, Goswami Y, Kumar A. Graphene–polyethylenedioxythiophene conducting polymer nanocomposite based supercapacitor. Electrochimica Acta. 2011; 56: 9406-9412.

[37] Kötz R, Carlen M. Principles and applications of electrochemical capacitors. Electrochimica acta. 2000; 45: 2483-2498.

[38] Zheng Y-z, Ding H-y, Zhang M-l. Preparation and electrochemical properties of nickel oxide as a supercapacitor electrode material. Materials Research Bulletin. 2009; 44: 403-407.

[39] Vidhyadharan B, Zain NKM, Misnon II, Abd Aziz R, Ismail J, Yusoff MM, Jose RJ. High performance supercapacitor electrodes from electrospun nickel oxide nanowires. Journal of Alloys Compounds. 2014; 610: 143-150.

[40] Kandalkar SG, Gunjakar J, Lokhande C. Preparation of cobalt oxide thin films and its use in supercapacitor application. Applied Surface Science. 2008; 254: 5540-5544.

[41] Xu J, Gao L, Cao J, Wang W, Chen Z. Preparation and electrochemical capacitance of cobalt oxide (Co3O4) nanotubes as supercapacitor material. Electrochimica Acta. 2010; 56: 732-736.

[42] Mitchell E, Gupta RK, Mensah-Darkwa K, Kumar D, Ramasamy K, Gupta BK, Kahol P. Facile synthesis and morphogenesis of superparamagnetic iron oxide nanoparticles for high-performance supercapacitor applications. New Journal of Chemistry. 2014; 38: 4344-4350.

[43] Kulal PM, Dubal DP, Lokhande CD, Fulari VJ. Chemical synthesis of Fe2O3 thin films for supercapacitor application. Journal of Alloys Compounds. 2011; 509: 2567-2571.

[44] Yu Z, Duong B, Abbitt D, Thomas J. Highly ordered MnO2 nanopillars for enhanced supercapacitor performance. Advanced materials. 2013; 25: 3302-3306.

[45] Wang P, Zhao Y-J, Wen L-X, Chen J-F, Lei Z-G. Ultrasound–microwave-assisted synthesis of MnO2 supercapacitor electrode materials. Industrial Engineering Chemistry Research. 2014; 53: 20116-20123.

[46] Lang X, Hirata A, Fujita T, Chen M. Nanoporous metal/oxide hybrid electrodes for electrochemical supercapacitors. Nature nanotechnology. 2011; 6: 232-236.

[47] Kuo S-L, Wu N-L. Investigation of pseudocapacitive charge-storage reaction of MnO2• nH2O supercapacitors in aqueous electrolytes. Journal of The Electrochemical Society. 2006; 153: A1317.

[48] Kozawa A, Powers R. The Manganese Dioxide Electrode in Alkaline Electrolyte; The Electron-Proton Mechanism for the Discharge Process from MnO2 to MnO1. 5. Journal of The Electrochemical Society. 1966; 113: 870.

[49] Lokhande V, Lokhande A, Lokhande C, Kim JH, Ji T. Super capacitive composite metal oxide electrodes formed with carbon, metal oxides and conducting polymers. J Journal of Alloys Compounds. 2016; 682: 381-403.

[50] Liao Q, Li N, Jin S, Yang G, Wang C. All-solid-state symmetric supercapacitor based on Co3O4 nanoparticles on vertically aligned graphene. ACS nano. 2015; 9: 5310-5317.

[51] Wang X-f, Ruan D-b, You Z. Pseudo-capacitive behavior of cobalt hydroxide/ carbon nanotubes composite prepared by cathodic deposition. Chinese journal of chemical physics. 2006; 19: 499.

[52] Ryu KS, Kim KM, Park N-G, Park YJ, Chang SH. Symmetric redox supercapacitor with conducting polyaniline electrodes. Journal of Power Sources. 2002; 103: 305-309.

[53] Gupta V, Miura N. High performance electrochemical supercapacitor from electrochemically synthesized nanostructured polyaniline. Materials Letters. 2006; 60: 1466-1469.

[54] Huang Y, Li H, Wang Z, Zhu M, Pei Z, Xue Q, Huang Y, Zhi C. Nanostructured polypyrrole as a flexible electrode material of supercapacitor. Nano Energy. 2016; 22: 422-438.

[55] Sharma R, Rastogi A, Desu S. Pulse polymerized polypyrrole electrodes for high energy density electrochemical supercapacitor. Electrochemistry Communications. 2008; 10: 268-272.

[56] Manjakkal L, Pullanchiyodan A, Yogeswaran N, Hosseini ES, Dahiya R. A wearable supercapacitor based on conductive PEDOT: PSS-coated cloth and a sweat electrolyte. Advanced Materials. 2020; 32: 1907254.

Chapter 5

Graphene Functionalization towards Developing Superior Supercapacitors Performance

Abd Elhamid M. Abd Elhamid, Heba Shawkey, Ahmed A.I. Khalil and Iftitan M. Azzouz

Abstract

Graphene is known as the miracle material of the 21st century for the wide band of participating applications and epic properties. Unlike the CVD monolayer graphene, Reduced graphene oxide (RGO) is a commercial form with mass production accessibility via numerous numbers of methods in preparation and reduction terms. Such RGO form showed exceptional combability in supercapacitors (SCs) where RGO is participated to promote flexibility, lifetime and performance. The chapter will illustrate 4 critical milestones of using graphene derivatives for achieving SC's superior performance. The first is using oxidized graphene (GO) blind with polymer for super dielectric spacer. The other three types are dealing with electrolytic SCs based on RGO. Polyaniline (PANI) was grown on GO for exceptionally stable SCs of 100% retention. Silver decoration of RGO was used for all-solid-state printable device. The solid-state gel electrolyte was developed by adding GO to promote current rating. Finally, laser reduced graphene is presented as a one-step and versatile technique for micropatterning processing. The RGO reduction was demonstrated from a laser GO interaction perspective according to two selected key parameters; wavelength and pulse duration.

Keywords: Polyaniline/RGO, Solid-state, laser-induced graphene, super dielectric spacer, flexible supercapacitor

1. Introduction

Supercapacitors (SCs) are a key block elements in our energy storage perspective that could stand alone or be combined with various types of batteries [1]. Unlike batteries, SCs possess unique features of high power and millions of cycling [2]. Yet, SCs energy density could not match the nowadays batteries [3], the development is focused on extending energy density and engineering flexible devices [2]. The storage mechanism is conducted to the SCs type as following; electric double-layer (EDL), pseudocapacitors and hybrid structure [4]. EDL is the simplest form of storing energy electrostatically due to electrolyte ions, whereas, pseudocapacitors is based on reversible redox reactions through active material's surface resulting in more than 10 times capacitance value than EDL. The hybrid type is combining faradic redox and non-faradic EDL reactions that showing near battery like performance.

Graphene is one of the most discussed electrodes material in energy storage due to outstanding electrical, mechanical and electrochemical performance [4]. Graphene functionalization in SCs is expected to lead the next SCs's generation processing as a result of the following; (i) high surface area of 2630 m^2/g correlated with low theoretical density of 2.28 g/cm^3, (ii) high carrier mobility and electrical conductivity that promote using of graphene as a compact active material/current collector, and (iii) Young's modulus of 1 TPa indicating excellent mechanical strength enabling perfect impeding in flexible and wearable electronics. The easy-to-get graphene form known as reduced graphene oxide (RGO) that is obtained from the reduction of highly oxidized exfoliated graphene oxide (GO). Crude RGO active material suffers from low specific capacitance and restacking over time that minimizes accessible surface area [5]. Therefore, many attempts of adding another material such as conducting polymers, metal oxides and metal NPS were reported [3, 6].

Despite a large number of graphene-based SCs reports, this chapter is focusing on selected milestones on using graphene in SCs according to extensive research work as well as others' reports. Four main points considered graphene derivatives, the first is using graphene oxide/polymer blind as a super dielectric spacer for double layer AC supercapacitor. The other technique is using RGO as efficient nucleation sites for polyaniline (PANI). After that, silver metal NPs decoration/RGO was used to fabricate flexible all-solid-state SCs of high power. Finally, laser-induced graphene is a one-step technique to obtain miniaturized 3D RGO based SCs.

2. Dielectric supercapacitor of pseudo 2D GO

Materials of high dielectric constant are directed for large capacitance circuit element, which enables minimizing dimensions in integrated circuits and basic storage elements, etc. [7–10]. Despite well-known ceramic materials such as Barium titanate-based composites, polymers are low cost and scalable but hold significantly low dielectric response. Thus, nano-composites doped inside a polymer matrix is a promising candidate for promoting dielectric characteristics for wearable and flexible electronics [5]. Among enormous compositions types, RGO and GO are blinded in polymer to expand the dielectric response via two different technique, the first is by forming multiple micro-capacitors inside the spacer while the other is using the oxide function group strong polarization [11]. The present section is revealing the potential of using graphene-based polymer composite potential as a super dielectric spacer for RF SC applications [12].

2.1 Double plate SC of PVA/GO spacer fabrication process

GO suspension of 1 gram in 0.1 L DI water and 8% PVA were used for preparing the mixture. The weight ratio blind was as following: 10%, 20% and 50% of GO to PVA using facile colloidal mixing method at 70°C. The resulted paste was coated on cleaned Al foils strips 0.08 × 40x50 mm^3 to form the compacted spacer. After mild evaporation, the top Al foil was attached to the three different ratios GO/PVA. The double-layer capacitor's spacer thickness was adjected to 475 μm after multiple drying processing using a hot press. The fabricated SCs electric characterization was conducted using LRC meter and multi-channels Potentiostat/Galvanostat. **Figure 1** illustrates a schematic for the whole study steps [12].

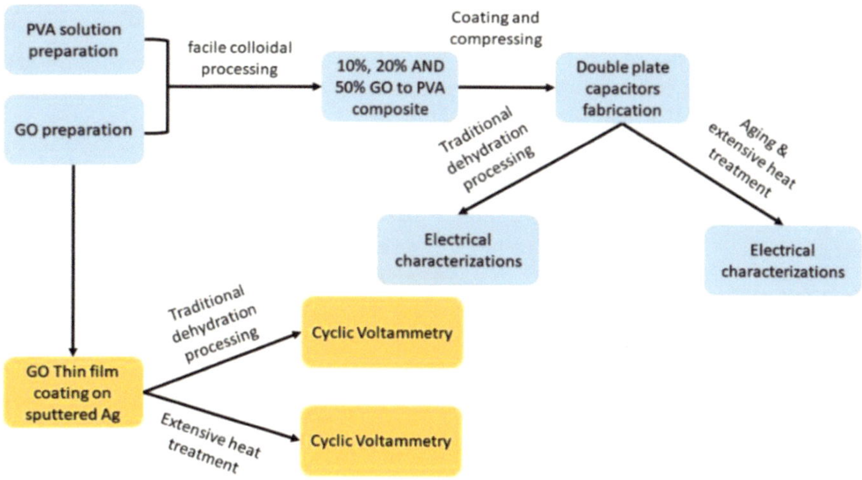

Figure 1.
Schematic of the GO/PVA and GO dielectric spacer study.

2.2 The mega dielectric value of double

The three different double plate capacitors of (10%, 20% and 50%) GO/PVA weight ratio are measured from 20 Hz to 1 MHz, the associated dielectric constants were estimated by knowing area and thickness of the spacers. The whole values are located in the range of mega value (10^6) as showed in **Figure 2(a)**. The dielectric decay over frequency is expected due to weak dipoles response. At low oscillating frequency, the corresponding oxide functionalized groups have the necessary time for alignment correlated with the applied electric field, which will promote the dielectric constant value due to strong polarization. Whereas, the increase in oscillating frequency will eliminate contributed ionic and space charge as well as cause a systematic drop in the dielectric response [13]. Thus, the GO filling inside the PVA polymer matrix achieved high dielectric SC caused by Maxwell- Wagner-Sillar theory [11]. The phase angle in **Figure 2(b)** is an indication of the leakage current through GO/PVA spacers. The higher phase angle than −90 degree resulted from free and bounded charges within the GO/PVA interface. Accordingly, SC's quality factor is linked to the measured phase angle. While the 10% gave the best spacer performance due to low water contents and residuals ions, the 50% ratio showed the lowest dielectric constant as well as quality factor. Low phase angle could be attributed to conductive graphitic defects generated during GO preparation.

The imaginary part (ε'') represents electric field dissipation into heat. The obtained losses could be attributed to GO lattice defects, conductive defects, water contents and ions residuals [13–15]. **Figure 2(c)** present the complex dielectric susceptibility curve (Cole-Cole), where ε is the complex permittivity, ε' real value and ε'' represent the imaginary part. Several GO oxide functional groups, defects, polymeric chain/GO interface and water molecules will result direct complex time response confirmed by Cole-Cole complex shapes. The 20% GO/PVA based SC cyclic voltammetry is presented in **Figure 2(d)**, which matched the previously measured mega dielectric constant value and performance. Those results indicate balanced filler loading within the polymer matrix of such 20% GO weight value [16]. Finally, the following **Table 1** reports the high dielectric value of using either crude GO or as a polymer matrix filler.

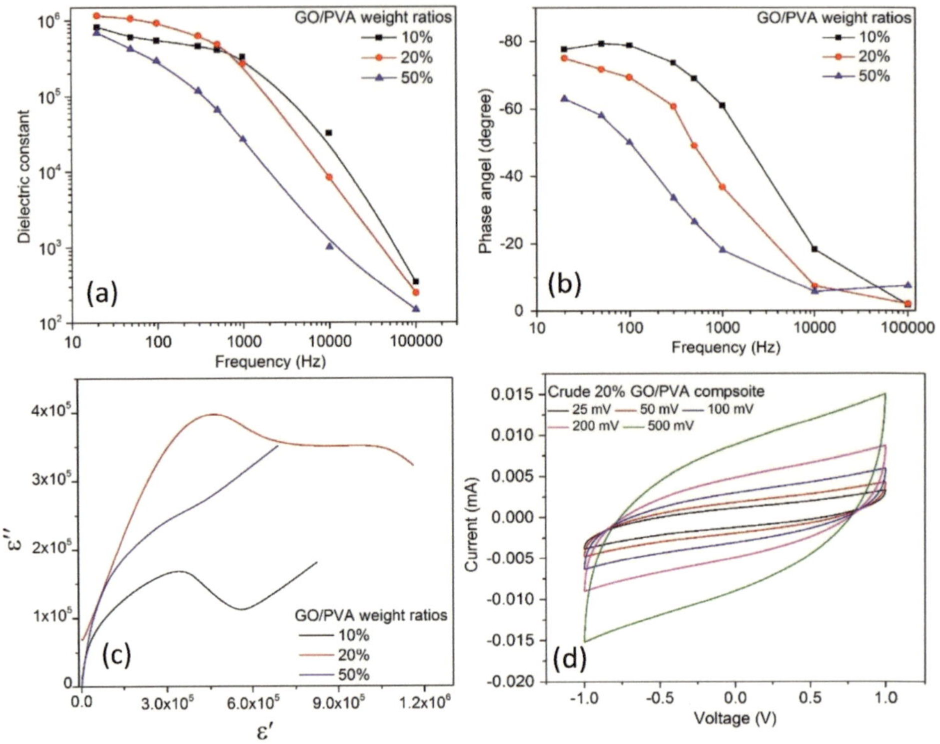

Figure 2.
The three GO/PVA ratio (a) dielectric constants, (b) phase angle, (c) Cole-Cole plot. And (d) the 20% GO/PVA ratio CV at different scan rates. Reproduced with permission from Ref. [12]. Copyright 2021, Elsevier Ltd.

Spacer Materials	DiC value	Reff.
This work (GO) (GO/PVA composite)	10^4 10^6	—
GO/PVA/polyethylene glycol	10^3	[17]
GO/PVA/Polypyrrole	10^3	[16]
GO/PVA/poly (4-styrene sulfonic acid)	10^3	[18]
GO foam	10^2	[14]
GO	pristine (10) annealing (10^3)	[19]
GO hybrid sponges	10^2 to 10^3	[20]
GO	10^4	[15]
GO	10^6	[13]

Table 1.
Dielectric constant of GO-based spacers. Reproduced with permission from Ref. [12]. Copyright 2021, Elsevier Ltd.

3. Conductive polymer graphene composite

3.1 Superior stable PANI reinforced RGO SC

An effective technique to not only prevent RGO restacking but also extend the electrochemical performance is surface composition via polymeric material [21]. PANI, polythiophene and polypyrrole are the most studied conductive polymers that participated in various application like SC electrodes, sensors, etc. [22]. Among those, PANI is preferable in SC applications due to low cost, controlled conductivity, easy processing and high electrochemical performance [23]. Unfortunately, poor stability is the main drawback. Hence, covalent grafting of PANI with a graphitic based material could promote lifetime, porosity and conductivity [24]. Oxidized graphene function groups are excellent nucleation sites for efficient covalent bonded PANI polymerization. Another advantage of using RGO/PANI composite in SC application is enabling the EDL behavior as well as pseudo-capacitance [25].

This section discusses PANI/RGO synthesis via two steps (i) *in situ* polymerization of distilled aniline on GO surface via initially adsorbed Fe^{2+} and (ii) reduction of the composite using hydrazine hydrate. The fabricated symmetric SCs was tested using four different electrolytes namely; sulfuric Acid, phosphoric acid. Potassium hydroxide and sodium sulfate to study the performance over a wide range of transported ions [26].

3.2 Preparation of PANI-RGO SC

2.25 g of Fe_2SO_4 was dissolved in 0.05 L DI water and was drop wised to 0.475 L of GO suspension (10 g/L) while stirred for 2 h. Aniline monomer was double distilled under vacuum for fresh using, and dissolved in 1 M (0.25 L HCl). The Fe_2SO_4 was rabidly added to GO suspension followed by adding Ammonium peroxydisulfate (APS) (4.5 g in 0.25 L 1 M of HCl and was kept stirring overnight at ambient conditions. After collecting and drying, the mixture was dissolved in 1 L DI water followed by sonication for reduction step. Hydrazine hydrate 1:1 to GO weight ratio was added in a boiling water bath. The electrodes PANI/RGO active material was conducted to another polymerization step before collected and drying in vacuum overnight. The paste was prepared by adding 8% PVDF to 92% PANI/RGO then coated 304 Stainless steel foil coated by sputtered 500 nm Pt current collector (sheet resistance is 2 Ω/□). **Figure 3** illustrates SC fabrication and characterization steps schematic [26].

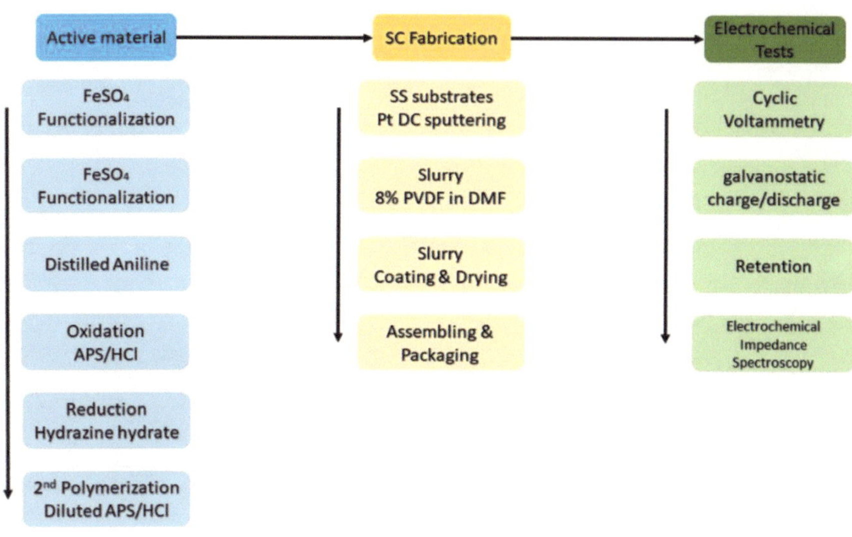

Figure 3.
Schematic of the PANI/RGO based SCs fabrication and measurements.

3.3 Plausible growth mechanism of PANI on RGO

The analysis of PANI polymerization, as well as GO reduction, was confirmed via standard characterization like XRD, Raman & EDX. However, the microscopic imaging is defining a more detailed view of folded RGO and PANI structure. SEM images are presented in **Figure 4(a)** and **(b)** that are showing the high load of PANI completely cover the RGO flakes in thick wood like shape and PANI/RGO flakes, respectively. TEM images illustrate RGO flake folded within dark region **Figure 4(c)**. Whereas, higher magnification TEM image present PANI growth on the RGO surface in tiny islands, which could be attributed to Fe^{2+} sites and GOs' function groups.

Dissociated $FeSO_4$ ions will be functionalized on the dispersed GO flakes. The distilled aniline monomer started to adsorbed on the GO surface [26]. By adding the APS/HCl, Fe (II) will be oxidized to Fe (III) and forming the major oxidation centers and the seed for PANI chains. The applied 1:1 weight ratio between aniline and GO will produce a high load of PANI correlated of high polymerization degree [27]. Fe (II)/APS is promoting effective and rapid polymerization due to direct bonding of the Fe (II) on pseudo-2D GO surface [27]. Thus, APS will react with Fe (II) ions not aniline due to low oxidation potential and produce sulfate radical anions (Eq. (1)) [28]:

$$Fe^{2+} + S_2O_8^{\ 2-} \rightarrow Fe^{3+} + SO_4^{\ 2-} + SO_4^{\ -\cdot} \quad (1)$$

Figure 4.
PANI-RGO at different magnification scale; (a) & (b) SEM images. (c) & (d) TEM images. Reproduced with permission from Ref. [26]. Copyright 2021, Elsevier Ltd.

3.4 Exceptionally stable PANI/RGO SC

The PANI/RGO based SCs were tested in various electrolytes to study the performance of the composite in strong alkaline, strong acid, weak acid and natural medium. Being a time-domain process, galvanostatic charge/discharge test is a

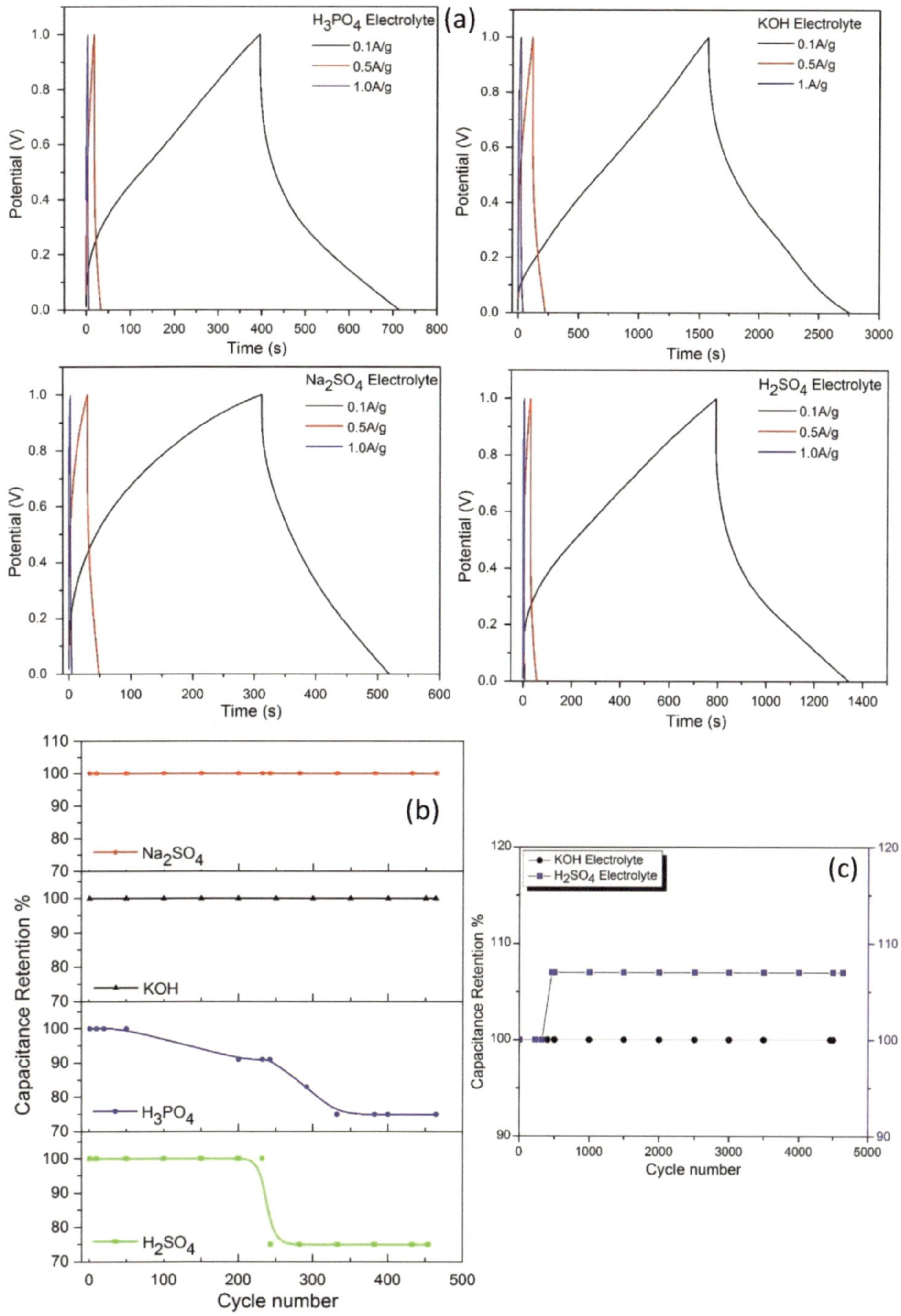

Figure 5.
Different electrolytes, (a) Galvanostatic charge/discharge curves cycle stabilities of different electrolytes. (b) Retention test for 460 cycles at 1A and (c) selected retention for 5000 cycles at 3A. Reproduced with permission from Ref. [26]. Copyright 2021, Elsevier Ltd.

versatile technique of defining specific capacitance, power, energy and lifetime. **Figure 5(a)** presents different electrolyte-based SC at different current. The estimated specific capacitance (0.1A) using H_2SO_4, H_3PO_4, Na_2SO_4 and KOH were 288, 126, 84 and 480 F/g, respectively. The capacitance of the composite could be expanded by using a highly conducting current collector, hot pressing of the active material, using standard cell configuration and a proper separator.

The remarkable result of the prepared PANI/RGO composite is the superior stability in various electrolytes as PANI is degradable at high temperature, high current and charging/discharging process. The first applied retention test at 1A was for nearly 500 cycles **Figure 5(b)**.

The pure EDL performance in alkaline KOH and natural Na_2SO_4 electrolytes present zero decay, whereas, sulfuric and phosphoric show ~75% retention due to aniline doping/degradation during charging. The 5000 cycles (**Figure 5(c)**) are considered to be relatively long cycling for a conducting polymer. The highest obtained capacitive results using H_2SO_4 and KOH were conducted to a 3A/g of current value for more aggressive cycling. Remarkably, KOH based SC maintained the 100% value correlated with the highest achieved capacitance. On the other hand, the sulfuric doping of PANI leads to an increase in the retention value behind 100% [29].

Table 2 presents the obtained results of PANI/RGO concerning others' reports comparatively to clear the role of RGO in enhancing the PANI super capacitive performance. It's worth mentioning that, 3-electrode setup multiplies the actual

Composite material/ Electrolyte	Specific capacitance (F/g)	No. of electrodes configuration	Retention	Process	Ref.
This work 6 M KOH 1 M H_2SO_4 1 M H_3PO_4	565 450 250	2	100% (5000) ~80% (5000) 75% (460)	Non faradic Faradic Non faradic	
PANI/GO/CNTs PVA/H_3PO_4 1 M H_2SO_4	89 729	2 3	80% (5000) 80% (5000)	Non faradic Faradic	[30]
G/PANI 1 M H_2SO_4	790	2	80% (5000)	Faradic	[31]
PANI/G/CNT 1 M H_2SO_4	510	3	91% (5000)	Faradic	[32]
PANI- N doped G/Pd 1 M H_2SO_4	230	2	96% (3000)	Faradic	[33]
PANI/G 1 M H_2SO_4	976	2	89.2% (1000)	Faradic	[34]
G/PANI 1 M H_2SO_4	529	3	85% (1000)	Faradic	[35]
PANI/G 1 M H_2SO_4	413	3	81% (1000)	Faradic	[36]
PANI/Carbon Qds/G 1 M H_2SO_4	871	3	72% (10000)	Faradic	[37]

Table 2.
The PANI-graphene (G) based SCs report. Reproduced with permission from Ref. [26]. Copyright 2021, Elsevier Ltd.

applied voltage that causes a massive increase in measured capacitance of about triple times when compared to the 2-electrode [38]. However, the 2-electrode cell configuration is preferable for practical SCs analysis. The electrochemical stability could be attributed to Fe(II) surface adsorbed on the GO and assisted APS strong locating oxidation.

4. Metal/RGO for solid-state and flexible structure

4.1 Printed devices based on doped graphene

Modern applications of wearable and flexible electronics require a convenient storage device to meet the rapid demands of energy consumption of devices such as ITO sensors and implantable bio-devices [39]. Long lifetime, high power profile, easy manufacturing and eco-friendly structure make the SCs is the best up-to-date candidate for powering such devices [40]. Nevertheless, solid and flexible SCs possess much lower power and energy densities than liquid-based devices that limit the practical coupling with systems. Accordingly, intense research work is applied for promoting the corresponding electrochemical characteristics. The electrodes active materials are already solid but the liquid electrolytes need sophisticated packaging that prevents planer designing. Ionic liquid salts, solid and electrolyte intercalated polymer are different types of the used solid electrolyte. In particular, gel polymer is dissociated ions from acid or base blinded in a polymer matrix. Solid-state gel polymer electrolytes present relatively good ionic conductivity, low cost, high stability, simple principle, reliable, environmentally friendly and safe to handle [41].

This part is using graphene to fabricated flexible, half printed and solid-state SC, the Ag decorated the RGO flakes to prevent restacking. The other functionalization of graphene oxide was for developing the gel polymer to promote high current and scan rate performance [42]. Despite the weak acidic nature of phosphoric acid, it was used as the main ions source inside the GP due to compatibility with RGO and stability over time other than strong acids like sulfuric.

4.2 SCs fabrication procedures

GO suspension was the start material (g/200 mL). Silver nitrate (1 g/50 mL) was added with two different volume ratios 4 ml and 16 ml to a 100 mL of GO suspension, respectively. The two ratios mixture were conducted to Hydrazine hydrate as a strong reduction agent and will be known as sample 1 (4 ml $AgNO_3$) while sample 2 (16 ml $AgNO_3$). The developed gel polymer was performed by dissolving 10% PVA at first and adding an equal weight of phosphoric acid. Finally, GO and PANI of weight 0.03 and 0.01 g were added to the GP, respectively.

A handmade plastic mask was used to make Ag/RGO electrode of dimensions 1x4 cm^2 (0.1 g) using doctor blade. After drying, the GP was coated followed by the second electrode. Lastly, the two devices were peeled off and a metal contact was sputtered on both sides. The schematic in **Figure 6**. illustrates devices fabrication procedures and applied electrochemical measurements [42].

4.3 The remarkable obtained printed SCs performance

TEM image in **Figure 7**. illustrates GO, RGO and Ag-doped RGO, respectively. The folded GO single layer is a clear sheet and the reduction of RGO make defects

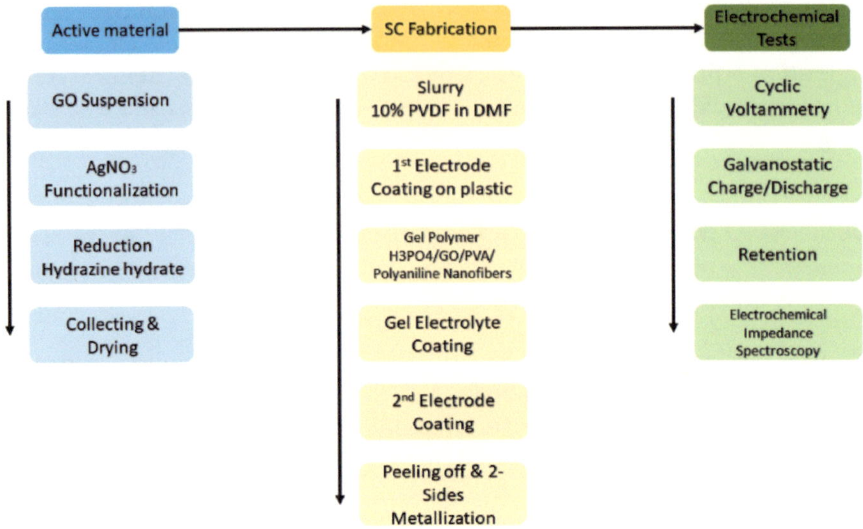

Figure 6.
Schematic of the Ag/RGO flexible SCs fabrication and measurements.

and multilayer represented by the dark regions. However, the silver nanoparticles incorporated RGO could be observed in the dark spots, the RGO do not present the dark defects and multilayered restacked shape due to the intercalated Ag particles. Thus, the as-prepared active material of Ag/RGO will show higher surface area, stability and conductivity. **Figure 7(d)** presents a captured image of the multi-layers printed symmetric SC on the support plastic film. And **Figure 7(e)** represents SC after peeling-off and current collectors metalization. Silver was used for metalization to reduce mismatch and reduce sheet resistance. The fabricated SCs were stretchable and flexible.

The Ag/RGO SCs was tested at a relatively high current value of 2A/g for a solid-state device by applying galvanostatic cycling. **Figure 8** shows the first three cycles where the time scale of sample 2 is nearly four-fold than sample 1 based SC. Sample 1 is located within the millisecond range, which confirms Ag doping ratio impact on such solid-state SC capacitive value. **Table 3** incorporates the estimated specific capacitance results of sample 1 and 2 via measured cycling voltammetry which is in good agreement with galvanostatic cycling results.

The electrochemical impedance spectroscopy is playing a key role in defining the solid electrolyte performance, the measurement was performed at a wide range from 0.1 Hz to 2 MHz **Figure 8(b)**. Nyquist plot and equivalent circuit measurements were obtained using VSP300 EC-Lab., the matched equivalent circuit is [R1 + C1/(R2 + Q2/(R3 + W3))], Q2 a constant phase element as a time-constant distribution function. R1 charge transfer resistance within electrodes. The low resistance region slop behavior is due to phosphoric weak acid natural [43]. Electrode/electrolyte interface is some sort of a junction represented by a small signal response equivalent circuit. The resistance originated from the charge transfer across such interfacial potential barrier besides electrolyte internal ohmic resistance. At extremely low frequency, the charge transfer resistance is polarization resistance. Whereas, the high frequency will reduce the resistance till reaching R1 value. R2 and R3 are the resistance of (silver current collector+ Ag/RGO active material+ metal/active material Interface contact). They are not equal as the peeling off cause roughness of one surface over the other that modified contact points and

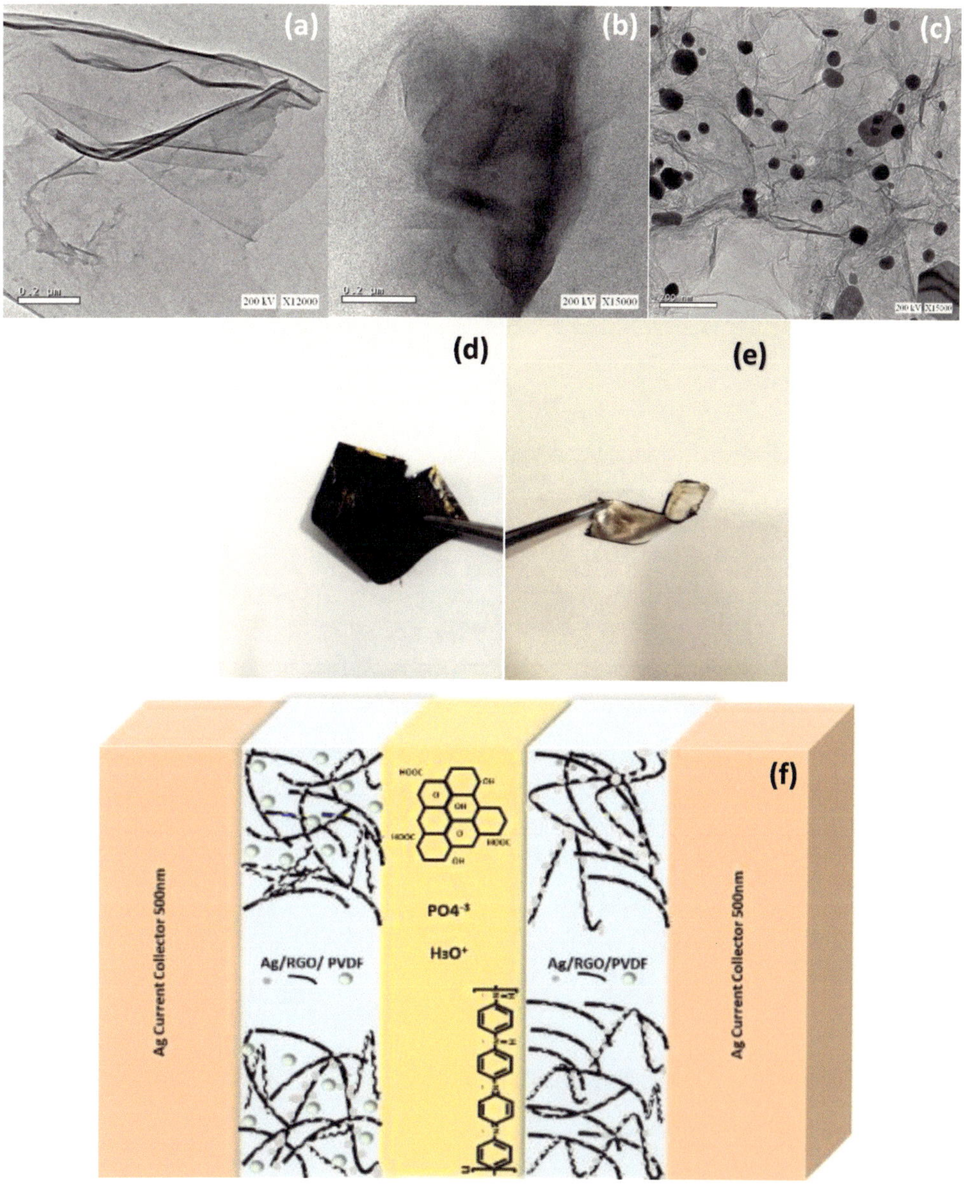

Figure 7.
TEM images of GO, RGO and Ag/RGO. Digital images of the flexible SCs. (d) on a plastic substrate. (e) after peeling and silver coating. (f) Final SC structure. Reproduced with permission from Ref. [42]. Copyright 2021, IEEE.

induced structural defects. During SC charging/discharging, Warburg impedance (W3) represnts semi-infinite one direction liner diffusion of electrolytic ions. The C1 is the magnitude of the two electrodes capacitance. The equivalent series resistance of sample 1 and sample 2 were 2.9 and 0.5317 ohms, respectively. Despite the obtained relatively small ESR for a solid-state SC, it could be further enhanced by the following; (i) using a high mobility ions based electrolyte, (ii) using a strong electrolyte, (iii) full printed structure using metallic ink as plasma deposition induce defects and (iv) applying layer by layer systematic hot compression and vaccum drying.

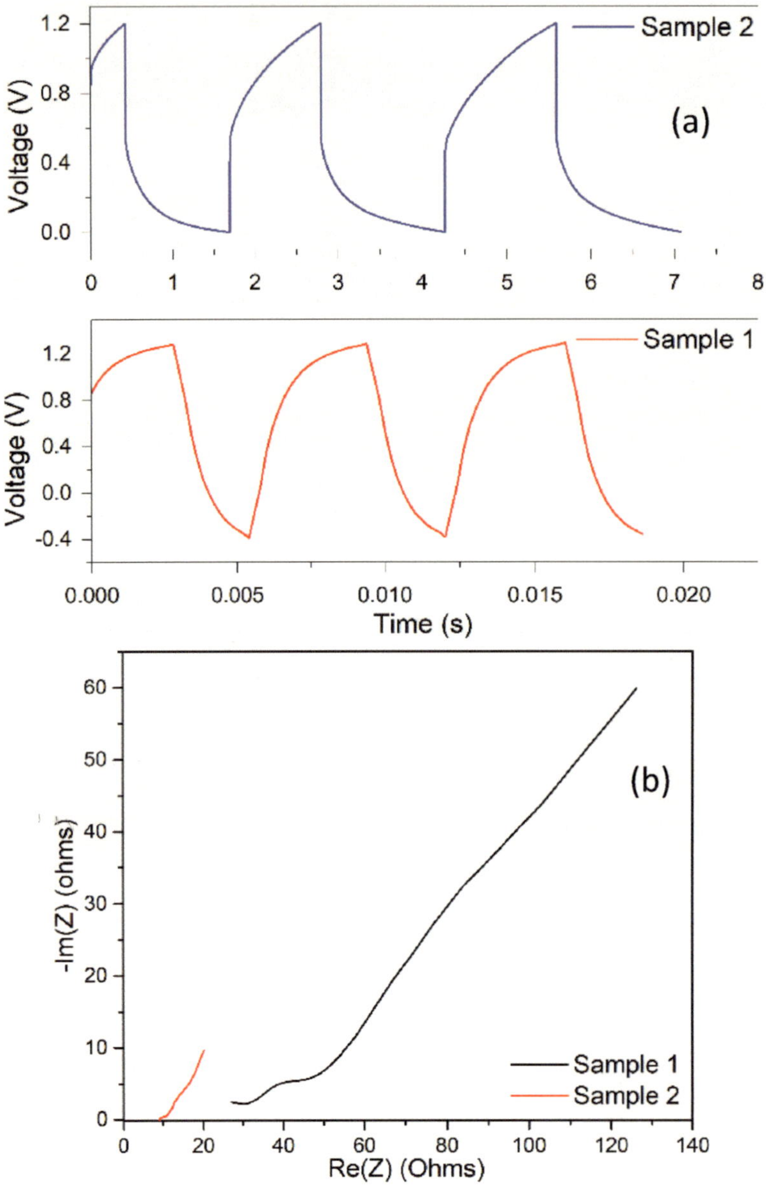

Figure 8.
(a) Charging/discharging test at 2A/g. (b) EIS test. Reproduced with permission from Ref. [42]. Copyright 2021, IEEE.

Scan rates	Sample 1	Sample 2
10 mv/sec	37.5 F/g	164 F/g
20 mv/sec	20 F/g	96.6 F/g
50 mv/sec	10 F/g	51.3 F/g
100 mv/sec	6 F/g	31.57 F/g
200 mv/sec	4.5 F/g	19.24 F/g

Table 3.
Estimated specific capacitance at different scan rates. Reproduced with permission from Ref. [42]. Copyright 2021, IEEE.

5. Laser-assisted RGO SC fabrication

5.1 3D RGO SC with laser irradiation

Conventional SCs are obeying two main configurations, the first is parallel plates and in-between a dielectric separator, the other is three parallel plates and in-between two dielectric separators to form two parallel-connected identical SCs. Those two designs are using a 3D structure and not convenient for planer, compact and miniaturized applications [44]. The developed IDE designs into interdigitated pattering of controlled dimensions will show diverse features over the conventional structure such as [45]; (i) fabrication on-plane macro/micro/nano in-plan SC, (ii) fast sidelong ions transportation, (iii) terminate separator usage and (iv) high power/energy due to confinement of electric field. For creating such fingers patterns SC, lithography techniques are used as it's a matter of certain dimensions etching processing. Photolithography, screen printing, laser dry etching and lift-off processing are used to fabricate complicated SCs designs [46]. However, most of SCs active material will

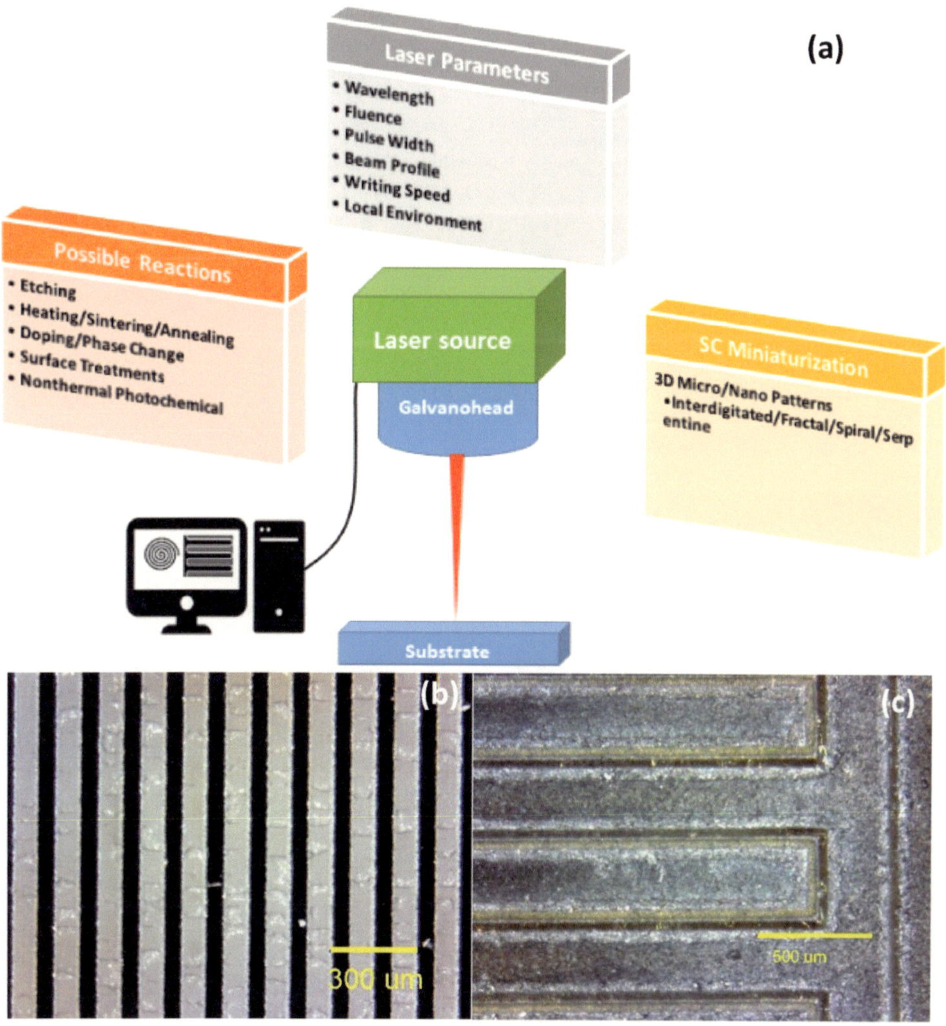

Figure 9.
(a) LDWT setup for SC patterning process. (b) and (c) microscopic images of ongoing using a nanosecond laser in micro SCs processing.

be affected by exposure to chemicals and multi-processing steps which will not match the easy-to-fabricated principle. Thus, laser processing can apply etching as well as controlled structure modification via laser material interaction.

GO is perfect a starting material for laser processing. Dispersion could be functionalized with various nanoparticles and deposited directly on substrates via liquid-phase processing [47]. The concept of using RGO as a commercial source of graphene especially in electronics applications is to reduce the oxygen groups to a highly conductive 2D layered structure. The previously mentioned flexible and PANI/RGO SC are conducted to chemical reduction techniques. However, there are various reduction techniques such as; thermal annealing, microwave flash reduction, laser-assisted processing and high-density UV reduction that affect the resulted RGO characteristics [48]. One of the effective techniques to reduce the dielectric GO into 3D network RGO for SC applications is laser direct writing

Laser	Description	Advantages	Drawbacks
CW CO_2 10600 nm	The most used laser in industry like cutting and welding. It can deliver direct heating to the GO film or carbonize polymer to obtain carbon like graphene Kapton.	• A relatively cheap laser system. • In vacuum or pure inert gas Ar/N_2 will promotes highly conductive RGO with an excellent degree of graphitization. • Mass production technique.	• Cannot apply etching or high precision patterning. • Complex multi-layer SC processing will not be affordable as there will be multi heat conduction constants and melting points. • Local heating will induce RGO defects.
CW 780, 788, 790 nm	It's one of the first trials in applying one-step GO reduction patterning at ambient conditions using CD lightscribe.	• Easy, cheap and results fair reduction.	• No control of power. • Cannot apply etching or high precision pattering. • Continuous beam induces RGO defects.
Pulsed nanosecond 1064/532/248 nm	This laser offers about Megawatt peak power which offers a unique interaction. Two main GO reduction process is working alternatively in major and minor possibility depend on the wavelength.	• Achieve reduction as well as etching processing of complex designs. • Fiber lasers enable mass production. • Could apply both photo (chemical/thermal) GO reduction.	• The thermal energy is propagating during etching within generating heat affected zone and affect the patterning precision.
Pulsed Femtosecond 800 nm	The Ultrafast laser is famous for cold nonthermal processing and pattering blow the diffraction limits.	• The generated pattern craters will be clean with well-defined edges. • Ability to create several nanometers pattern and spacing SC in just one step.	• The strong cold processing in not effective in restoring graphene-like structure and related electric properties.

Table 4.
Different laser type for SCs fabrication process.

technique (LDWT) that enables fast, low cost, easy-processing and maskless patterning [46]. Laser processing of SCs is accessible for most of available rigid and flexible substrates.

5.2 LDWT of RGO based miniaturized SC

Figure 9 presents the LDWT processing main parameters that affect the resulted SC performance. Many reports discuss the corresponding laser parameters, interactions and resulted RGO characteristics [49, 50]. During laser processing, GO reduction is governed by photothermal oxide group removal and photochemical bond breaking. Laser wavelength and pulse duration are the two main key parameters of SC processing. The laser wavelength controls the major photoreduction process of being photochemical (≤ 400 nm) or photothermal (≥ 400 nm).

photochemical process is directly conducted to photon energy that can break oxygen functional groups. The photothermal process is focused on near-infrared region and infrared bands. The deposited thermal energy on GO is maximized upon increasing the beam density to induce local heating within a certain heat-affected zone. This sort of heat can break oxygen radicals as well. In addition, long wavelengths of fast/ultrafast pulse duration could apply photochemical reduction through nonlinear processes and multiphoton absorption [51]. **Table 4** is concluding the GO reduction process using most common laser sources [52–54].

6. Conclusions

Finally, graphene not only proved effective participation in different types of SCs but showing superior electrochemical and electromagnetic performance. PVA/GO composite could work as an efficient super dielectric spacer of mega value 10^6 thanks to the intercalated water dipoles and oxide functional groups. Maxwell- Wagner- Sillar effect expanded the PVA blind dielectric response. By using Fe(II) as a growth mediated oxidizer for PANI growth on GO surface, 100% retention at high current after 5000 cycles at 3A/g was obtained. The PANI/RGO showed EDL performance for H_3PO4, Na_2SO_4 and KOH. For a flexible and printed SC fabrication, silver doped RGO has 29.5 Wh/Kg energy at 2A. The high current rating was promoted via GO impeding inside the electrolytic solid polymer matrix. The miniaturization of one-step fabricated SC is using laser-assisted reduction technique. The CO_2 is a cost-effective direct heating source that induces thermal reduction, whereas, a femtosecond laser is using cold processing that reduces the oxide groups but with a low degree of graphenezation. Nanosecond laser sources combine thermal and photochemical reduction through the GO film.

Conflict of interest

We declare no conflict of interest.

Author details

Abd Elhamid M. Abd Elhamid[1,2]*, Heba Shawkey[3], Ahmed A.I. Khalil[1] and Iftitan M. Azzouz[1]

1 Laser Sciences and Interactions Department, National Institute of Laser Enhanced Sciences (NILES), Cairo University, Giza, Egypt

2 Nanotechnology Laboratory, Electronics Research Institute, El Nozha, Egypt

3 Microelectronics Department, Electronics Research Institute, El Nozha, Egypt

*Address all correspondence to: a.m.abdelhamid@eri.sci.eg

References

[1] Capasso C, Veneri O. Integration between Super-capacitors and ZEBRA Batteries as High Performance Hybrid Storage System for Electric Vehicles. Energy Procedia. 2017;105:2539-44. DOI: 10.1016/j.egypro.2017.03.727

[2] Huang S, Zhu X, Sarkar S, Zhao Y. Challenges and opportunities for supercapacitors. APL Mater. 2019;7. DOI: 10.1063/1.5116146

[3] Majumdar D, Mandal M, Bhattacharya SK. Journey from supercapacitors to supercapatteries: recent advancements in electrochemical energy storage systems. Emergent Mater. 2020;3:347-67.

[4] Banerjee S, Sinha P, Verma KD, Pal T, De B, Cherusseri J, et al. Capacitor to Supercapacitor BT - Handbook of Nanocomposite Supercapacitor Materials I: Characteristics. In: Kar KK, editor. Cham: Springer International Publishing; 2020. p. 53-89. DOI: 10.1007/978-3-030-43009-2_2

[5] El-Gendy DM, Ghany NA, El Sherbini EF, Allam NK. Adenine-functionalized Spongy Graphene for Green and High-Performance Supercapacitors. Sci Rep. 2017;7:1-10. DOI: 10.1038/srep43104

[6] Li Q, Horn M, Wang Y, MacLeod J, Motta N, Liu J. A review of supercapacitors based on graphene and redox-active organic materials. Materials; 2019;12: 703. DOI: 10.3390/ma12050703

[7] Lu CY, Chang-Liao KS, Lu CC, Tsai PH, Kyi YY, Wang TK. Investigation of voltage-swing effect and trap generation in high-k gate dielectric of MOS devices by charge-pumping measurement. Microelectron Eng. 2008;85:20-6. DOI: 10.1016/j.mee.2007.02.012

[8] Wilk RW. High dielectric constant materials. In: Gilmer, D. Huff HR, editors. Springer; 2005. 707 p. DOI: 10.1007/b137574

[9] Behera P, Ravi S. Effect of Ni doping on structural, magnetic and dielectric properties of M-type barium hexaferrite. Solid State Sci. 2019;89:139-149. DOI: 10.1016/j.solidstatesciences.2019.01.003

[10] Das R, Choudhary RP. Studies of structural, dielectric relaxor and electrical characteristics of lead-free double Perovskite: Gd2NiMnO6. Solid State Sci. 2019;87:1-8. DOI: 10.1016/j.solidstatesciences.2018.10.020

[11] Wadhwa H, Kandhol G, Deshpande UP, Mahendia S, Kumar S. Thermal stability and dielectric relaxation behavior of in situ prepared poly(vinyl alcohol) (PVA)-reduced graphene oxide (RGO) composites. Colloid Polym Sci. 2020;298:1319-33. DOI: 10.1007/s00396-020-04718-0

[12] Abd Elhamid AM, Shawkey H, Nada AA, Bechelany M. Anomalous dielectric constant value of graphene oxide/Polyvinyl alcohol thin film. Solid State Sci. 2019;94:28-34.

[13] Kumar KS, Pittala S, Sanyadanam S, Paik P. A new single/few-layered graphene oxide with a high dielectric constant of 106: contribution of defects and functional groups. RSC Adv. 2015;5:14768-79. DOI: 10.1039/C4RA10800K

[14] Hou ZL, Liu X Da, Song WL, Fang HM, Bi S. Graphene oxide foams: the simplest carbon-air prototypes for unique variable dielectrics. J Mater Chem C. 2017;5:3397-407. DOI: 10.1039/c6tc04971k

[15] Liu J, Galpaya D, Notarianni M, Yan C, Motta N. Graphene-based thin

film supercapacitor with graphene oxide as dielectric spacer. Appl Phys Lett. 2013;103:1-5. DOI: 10.1063/1.4818337

[16] Deshmukh K, Ahamed MB, Pasha SKK, Deshmukh RR, Bhagat PR. Highly dispersible graphene oxide reinforced polypyrrole/polyvinyl alcohol blend nanocomposites with high dielectric constant and low dielectric loss. RSC Adv. 2015;5:61933-45. DOI: 10.1039/C5RA11242G

[17] Deshmukh K, Ahamed MB, Sadasivuni KK, Ponnamma D, Deshmukh RR, Pasha SKK, et al. Graphene oxide reinforced polyvinyl alcohol/polyethylene glycol blend composites as high-performance dielectric material. J Polym Res. 2016;23:159. DOI: 10.1007/s10965-016-1056-8

[18] Deshmukh K, Ahamed MB, Sadasivuni KK, Ponnamma D, AlMaadeed MAA, Khadheer Pasha SK, et al. Graphene oxide reinforced poly (4-styrenesulfonic acid)/polyvinyl alcohol blend composites with enhanced dielectric properties for portable and flexible electronics. Mater Chem Phys. 2017;186:188-201. DOI: 10.1016/j.matchemphys.2016.10.044

[19] Kavinkumar T, Sastikumar D, Manivannan S. Effect of functional groups on dielectric, optical gas sensing properties of graphene oxide and reduced graphene oxide at room temperature. RSC Adv. 2015;5:10816-25. DOI: 10.1039/C4RA12766H

[20] Wang Y, Zhang KL, Zhang BX, Ma CJ, Song WL, Hou ZL, et al. Smart mechano-hydro-dielectric coupled hybrid sponges for multifunctional sensors. Sensors Actuators, B Chem. 2018;270:239-46. DOI: 10.1016/j.snb.2018.05.023

[21] Radhakrishnan S, Kim SJ. Facile fabrication of NiS and a reduced graphene oxide hybrid film for nonenzymatic detection of glucose. RSC Adv. 2015;5:44346-52. DOI: 10.1039/c5ra01074h

[22] Guo X, Facchetti A. The journey of conducting polymers from discovery to application. Nature Materials. Nature Research; 2020;19:922-8. DOI: 10.1038/s41563-020-0778-5

[23] S. Radhakrishnan, C. Rao MV. Performance of Conducting Polyaniline-DBSA and Polyaniline-DBSA/Fe3O4 Composites as Electrode Materials for Aqueous Redox Supercapacitors. J Appl Polym Sci. 2011;122:1510-8. DOI: 10.1002/app.34236

[24] Li ZF, Zhang H, Liu Q, Liu Y, Stanciu L, Xie J. Covalently-grafted polyaniline on graphene oxide sheets for high performance electrochemical supercapacitors. Carbon. 2014;71:257-67. DOI: 10.1016/j.carbon.2014.01.037

[25] Eftekhari A, Li L, Yang Y. Polyaniline supercapacitors. J Power Sources. 2017;347:86-107. DOI: 10.1016/j.jpowsour.2017.02.054

[26] Alamin AA, Abd Elhamid AM, Anis WR, Attiya AM. Fabrication of symmetric supercapacitor based on relatively long lifetime polyaniline grown on reduced graphene oxide via Fe2+ oxidation sites. Diam Relat Mater. 2019;96:182-94.

[27] Li G, Zhang C, Li Y, Peng H, Chen K. Rapid polymerization initiated by redox initiator for the synthesis of polyaniline nanofibers. Polymer. 2010;51:1934-9. DOI: 10.1016/j.polymer.2010.03.004

[28] Odian G. Principles of polymerization. 4th ed. New Jersey: John Wiley & Sons; 2004. p. 216. DOI: 10.1002/047147875X

[29] Tawde S, Mukesh D, Yakhmi J V. Redox behavior of polyaniline as influenced by aromatic sulphonate

anions: Cyclic voltammetry and molecular modeling. Synth Met. 2002;125:401-13. DOI: 10.1016/S0379-6779(01)00483-0

[30] Jiang Q, Shang Y, Sun Y, Yang Y, Hou S, Zhang Y, et al. Flexible and multi-form solid-state supercapacitors based on polyaniline/graphene oxide/CNT composite films and fibers. Diam Relat Mater. 2019;92:198-207. DOI: 10.1016/j.diamond.2019.01.004

[31] P. Yu, X. Zhao, Z. Huang YL and QZ. Free-standing three-dimensional graphene and polyaniline nanowire arrays hybrid foams for high-performance flexible and lightweight supercapacitors. J Mater Chem A. 2014;4:14413-7. DOI: 10.1039/C4TA02721C

[32] Li D, Li Y, Feng Y, Hu W, Feng W. Hierarchical graphene oxide/polyaniline nanocomposites prepared by interfacial electrochemical polymerization for flexible solid-state supercapacitors. J Mater Chem A. 2015;3:2135-43. DOI: 10.1039/c4ta05643d

[33] Kalambate PK, Rawool CR, Karna SP, Srivastava AK. Nitrogen-doped graphene/palladium nanoparticles/porous polyaniline ternary composite as an efficient electrode material for high performance supercapacitor. Mater Sci Energy Technol. 2018;2:246-57. DOI: 10.1016/j.mset.2018.12.005

[34] He S, Hu X, Chen S, Hu H, Hanif M, Hou H. Needle-like polyaniline nanowires on graphite nanofibers: Hierarchical micro/nano-architecture for high performance supercapacitors. J Mater Chem. 2012;22:5114-20. DOI: 10.1039/c2jm15668g

[35] Dai W, Ma L, Gan M, Wang S, Sun X, Wang H, et al. Fabrication of sandwich nanostructure graphene/polyaniline hollow spheres composite and its applications as electrode materials for supercapacitor. Mater Res Bull. 2016;76:344-52. DOI: 10.1016/j.materresbull.2015.12.045

[36] Ning G, Li T, Yan J, Xu C, Wei T, Fan Z. Three-dimensional hybrid materials of fish scale-like polyaniline nanosheet arrays on graphene oxide and carbon nanotube for high-performance ultracapacitors. Carbon. 2013;54:241-8. DOI: 10.1016/j.carbon.2012.11.035

[37] S. Li, Ai. Gao, F.Yi, D. Shu, H. Cheng, Xi Zhou, C. He DZ and FZ. Preparation of carbon dots decorated graphene/polyaniline composites by supramolecular in-situ self-assembly for high-performance supercapacitors. Electrochim Acta. 2019;297:1094-103. DOI: 10.1016/j.electacta.2018.12.036

[38] Stoller MD, Ruoff RS. Best practice methods for determining an electrode material's performance for ultracapacitors. Energy Environ Sci. 2010;3:1294-301. DOI: 10.1039/c0ee00074d

[39] Railanmaa A, Kujala M, Keskinen J, Kololuoma T, Lupo D. Highly flexible and non-toxic natural polymer gel electrolyte for printed supercapacitors for IoT. Appl Phys A Mater Sci Process. 2019;125:168. DOI: 10.1007/s00339-019-2461-8

[40] Jian X, Yang H min, Li J gang, Zhang E hui, Cao L le, Liang Z hai. Flexible all-solid-state high-performance supercapacitor based on electrochemically synthesized carbon quantum dots/polypyrrole composite electrode. Electrochim Acta. 2017;228:483-93. DOI: 10.1016/j.electacta.2017.01.082

[41] Alipoori S, Mazinani S, Aboutalebi SH, Sharif F. Review of PVA-based gel polymer electrolytes in flexible solid-state supercapacitors: Opportunities and challenges. Journal of Energy Storage. 2020;27:101072. DOI: 10.1016/j.est.2019.101072

[42] Elhamid AM, Alamin AA, Selim AM, Wasfey MA, Zahran MB. Fabrication of Flexible, Half printed and All-Solid-State Symmetric Supercapacitor Based on Silver Decorated Reduced Graphene Oxide. In: ACCS/PEIT 2019 - 2019 6th International Conference on Advanced Control Circuits and Systems; Hurgada:IEEE: 2019. p. 151-5. DOI: 10.1109/ACCS-PEIT48329.2019.9062877

[43] Chen Q, Li X, Zang X, Cao Y, He Y, et al. Effect of different gel electrolytes on graphene- based solid-state supercapacitors. RSC Adv. 2014;36253:36253-6. DOI: 10.1039/c4ra05553e

[44] Yu C, An J, Chen Q, Zhou J, Huang W, Kim Y, et al. Recent Advances in Design of Flexible Electrodes for Miniaturized Supercapacitors. Small Methods. 2020;4:1900824. DOI: 10.1002/smtd.201900824

[45] Liu N, Gao Y. Recent Progress in Micro-Supercapacitors with In-Plane Interdigital Electrode Architecture. Small. 2017;13:1701989. DOI: 10.1002/smll.201701989

[46] Velasco A, Ryu YK, Boscá A, Ladrón-De-Guevara A, Hunt E, Zuo J, et al. Recent trends in graphene supercapacitors: From large area to microsupercapacitors Sustainable Energy and Fuels. 2021;5:1235-54. DOI: 10.1039/d0se01849j

[47] Ahmad H, Fan M, Hui D. Graphene oxide incorporated functional materials: A review. Coordination Chemistry Reviews. 2018;145:270-80. DOI: 10.1016/j.ccr.2017.03.021

[48] Kuila T, Mishra AK, Khanra P, Kim NH, Lee JH. Recent advances in the efficient reduction of graphene oxide and its application as energy storage electrode materials Nanoscale. 2013:5:52-71. DOI: 10.1039/c2nr32703a

[49] Kumar R, Singh RK, Singh DP, Joanni E, Yadav RM, Moshkalev SA. Laser-assisted synthesis, reduction and micro-patterning of graphene: Recent progress and applications. Coordination Chemistry Reviews. 2017;342:34-79. DOI: 10.1016/j.ccr.2017.03.021

[50] Thekkekara L V. Direct Laser Writing of Supercapacitors. In: Liudvinavičius L, editor. Supercapacitors - Theoretical and Practical Solutions. InTech; 2018. p. 103-16. DOI: 10.5772/intechopen.73000

[51] Okhrimchuk AG, Mezentsev VK, Schmitz H, Dubov M, Bennion I. Cascaded nonlinear absorption of femtosecond laser pulses in dielectrics. Laser Phys. 2009;19:1415-22. DOI: 10.1134/s1054660x09070081

[52] Arul R, Oosterbeek RN, Robertson J, Xu G, Jin J, Simpson MC. The mechanism of direct laser writing of graphene features into graphene oxide films involves photoreduction and thermally assisted structural rearrangement. Carbon. 2016;99:423-31. DOI: 10.1016/j.carbon.2015.12.038

[53] de Lima BS, Bernardi MIB, Mastelaro VR. Wavelength effect of ns-pulsed radiation on the reduction of graphene oxide. Appl Surf Sci. 2020;506:144808. DOI: 10.1016/j.apsusc.2019.144808

[54] Bhattacharjya D, Kim CH, Kim JH, You IK, In J Bin, Lee SM. Fast and controllable reduction of graphene oxide by low-cost CO 2 laser for supercapacitor application. Appl Surf Sci. 2018;462:353-61. DOI: 10.1016/j.apsusc.2018.08.089

Section 4

MXene Based Materials

Chapter 6

Two-Dimensional MXene Based Materials for Micro-Supercapacitors

Aditya Sharma and Chandra Sekhar Rout

Abstract

With the boom in the development of micro-electronics for wearable and flexible electronics, there is a growing demand for micro-batteries and micro-supercapacitors (MSCs). Micro-supercapacitors have garnered a considerable attention for the evolution of these energy storage micro-systems. The choice of electrode material plays a pivotal role in the fabrication and development of MSCs. Recently, a new emerging family of two-dimensional transition metal (M) carbides or nitrides (X) cited as 2D MXene has emerged as a novel material. Due to its exceptionally high electronic conductivity ~10,000 S cm^{-1}, high charge storage capacity and easy processing capability helps to use MXene as the promising candidate for micro-supercapacitors electrodes. Taking the advantage of such exceptional properties. MXenes have been explored enormously in stacked as well as in interdigital architecture for on-chip micro-supercapacitors (MSCs). This book chapter includes a recent advancement of MXene based MSCs, with a brief overview of synthesis and fabrication techniques.

Keywords: 2D MXene, Micro-supercapacitor, Wearable and flexible electronics, Energy storage, Micropattern

1. Introduction

There is an increase in demand for flexible and solid-state on chip micro-electronics for smart wearable micro-devices for energy, environmental, biological, medical and various other applications which can be either wireless or integrated with solar or piezoelectric energy harvesters. Great efforts have been made by the scientists to design and develop smart as well as portable microsystems, primarily for self-powered and on-chip integrated power systems. To cope with the increasing demand of micro-electronics, there is an abrupt rise in the demand of micro-energy storage devices. However, micro-batteries are restricted due to their limited life and power density. Micro-supercapacitors (MSCs) hold the best alternate to the micro-batteries, despite the lower energy density. In contrast, MSCs can demonstrate superior cycle life, faster charge/discharge rates, high power density as well as overall stable performance which is promising [1]. Presently, MSCs have two types of architecture, one with the conventional sandwich type and others are in-plane interdigital pattern type as shown in **Figure 1(a, b)**, [2]. Generally, the interdigitated coplanar design offers better performance due to the short ion diffusion distance which gives the enormous surface area. Thus, exhibiting an excellent rate capability, high-power density and ease of integration with micro-devices [1, 4, 5].

MXene family, exhibiting a high electronic conductivity up to~2.4×10^4 S/cm and volumetric capacitance 1500 F/cm^3 with good rate capability of 10 V/s in acidic electrolyte [5]. Hence there is plenty of room to design and develop MXene based Micro-supercapacitor devices [20]. The 2D nature, excellent mechanical stability and exceptionally tunable physio-chemical properties makes MXenes, the best candidate for MSC device. This book chapter includes various direct–indirect techniques to fabricate MXene based MSC device.

2. MXenes: brief review

2.1 Synthesis

There are generally two methods to synthesize MXene i.e., (1) top-down approach which includes selective etching or exfoliation of metal layer and (2) bottom-up approach including chemical vapor deposition (CVD), template assisted growth method. Wet chemical etching i.e., fluoride based acids are most commonly reported methods to etch "A" element, generally group IIIA and IVA group (Al or Si) elements from MAX phases (one or several atomic layers) which are replaced by functional groups, where M is termed as early transition metals, from group IIIB to IVB, and X is carbide/nitride by using different wt % of fluoride containing acid such as HF or mixture of LiF-HCl acid [17, 20, 21]. The first ever report to synthesis MXene by eliminating the aluminium layer from Ti_3AlC_2 (MAX) by using hydrogen fluoride (HF) in the range of 10 to 50 % concentration of etchant [17]. The exfoliated 2D $Ti_3C_2T_x$ possess excellent 2D sheets like morphology almost similar to graphene sheets as shown in **Figure 1(c,d)** [3]. To avoid highly acidic HF acid, various other methods have been developed to produce in-situ HF salts comparatively less hazardous than HF. Recently, a new approach to etch with molten salts allows to dissolve A-element at high temperature [21]. This method demonstrated the complete removal of Fluorine ions and found to be much purer MXene than one etched with only HF. The presence of surface functional groups like -OH, -F and -O etc. improves the hydrophilic character of MXenes which further enhances the stability. The reaction mechanism of firstly synthesized $Ti_3C_2T_X$ (MXene) by etching Al layers from Ti_3AlC_2 (MAX phase) with HF is given in **Figure 2(a)** [22].

$$Ti_3AlC_2 + 3\mathrm{HF} = AlF_3 + 3/2\,\mathrm{H_2} + Ti_3\mathrm{C_2} \tag{1}$$

$$Ti_3\mathrm{C_2} + 2\mathrm{H_2O} = Ti_3\mathrm{C_2}(OH)_2 + \mathrm{H_2} \tag{2}$$

$$Ti_3\mathrm{C_2} + 2\mathrm{HF} = Ti_3\mathrm{C_2F_2} + \mathrm{H_2} \tag{3}$$

Reaction (2) and (3) gives rise to the surface terminations like -O, -OH, -F etc., respectively [17, 24]. MXenes have three possible structures with different layers of stacking as shown in **Figure 2(b)**, [22]. The tentative elements of MXene precursor in the periodic table predicted till now presented in **Figure 2(c)** [23].

The timeline of synthesis of MXene in different year is given in **Figure 3(a)**. [25] Choice of synthesis and processing method including precursors etchant, intercalant, reaction sonication time etc. strongly influence the properties of resultant MXene. Mild alkali etchants like NaOH [26] and $NaBF_4$ [27] were also proposed to synthesis Ti_3C_2 by high temperature hydrothermal etching of Al layer from Ti_3AlC_2. This method extended to other MXenes such as Nb_2C [27]. Similarly, molten $ZnCl_2$ were

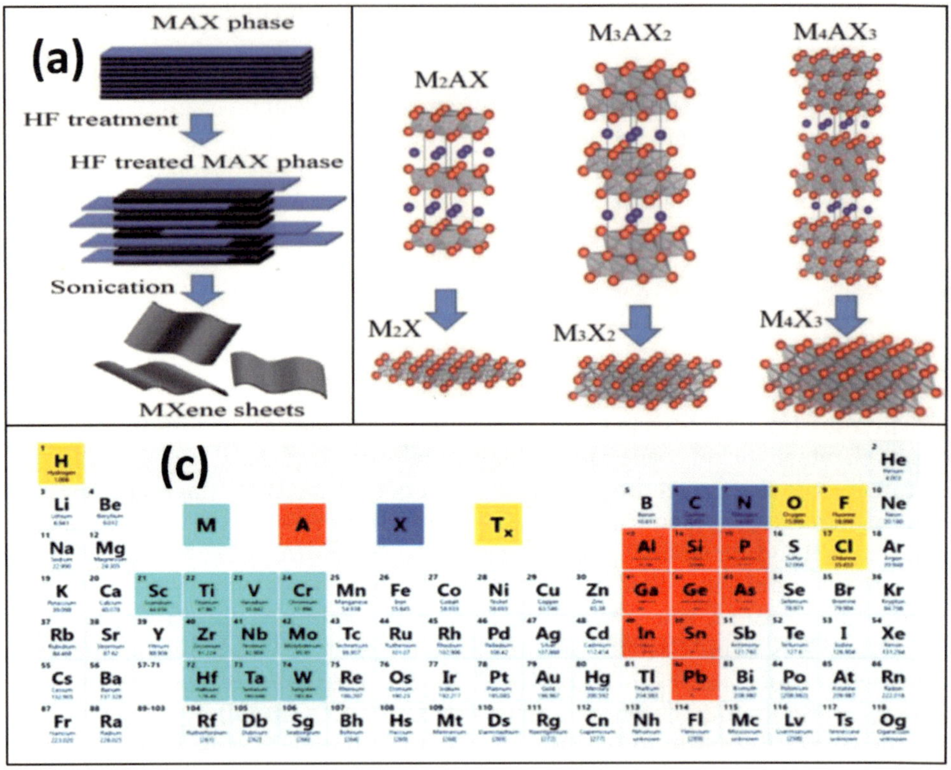

Figure 2.
(a) Schematic showing the synthesis of MXene from MAX by HF treatment, (b) synthesis mechanism of different order of MXenes by MAX phases. [22] and (c) compositions of MXene elements in periodic table (Reprinted from [23] with permission from, copyright@2020, MDPI).

used for different MAX phases like Ti_3AlC_2, Ti_2ZnN, Ti_2AlC and V_2AlC to substitute Zn^{2+} ions [28]. Another report suggested a fluoride free-electrochemical etching at room temperature synthesis of Ti_2C and Ti_3C_2 in dilute HCl [29] and NH_4Cl/TMAOH [30]. Also, a naturally delaminated MXenes with better electronic conductivity can be produced using minimally intensive layer delamination (MILD) without the use of further handshaking or sonication of MXenes [25]. More than 20 different types of MXenes have been synthesized experimentally [18]. **Figure 3(b)** represents the different etching methods that are used to synthesis different MXene products [25].

2.2 Properties

2.2.1 Electrical conductivity

MXenes have been extensively investigated by computational methods [31, 32], MXenes can be categorized into three types i.e., metallic, semi-metallic and semi-conducting [33]. Generally, bare MXenes have very high electronic conductivity with high density of states (DOS) at the fermi level. Electronic properties of MXenes are strongly influenced by the surface, morphology and stacking behavior of MXene sheets. Delaminated MXenes flakes show ultra-high electronic conductivity of upto 9880 Scm^{-1}, which can be further tuned by modifying the surface-terminations [34]. In addition to this, MXenes strongly depends on synthesis procedure which can be achieved by varying the synthesis conditions. HF etched highly defective MXene exhibits electronic conductivity of 1,000 Scm^{-1}. Whereas it improved to 4600 Scm^{-1} for powder and further enhanced to 6500 Scm^{-1} for

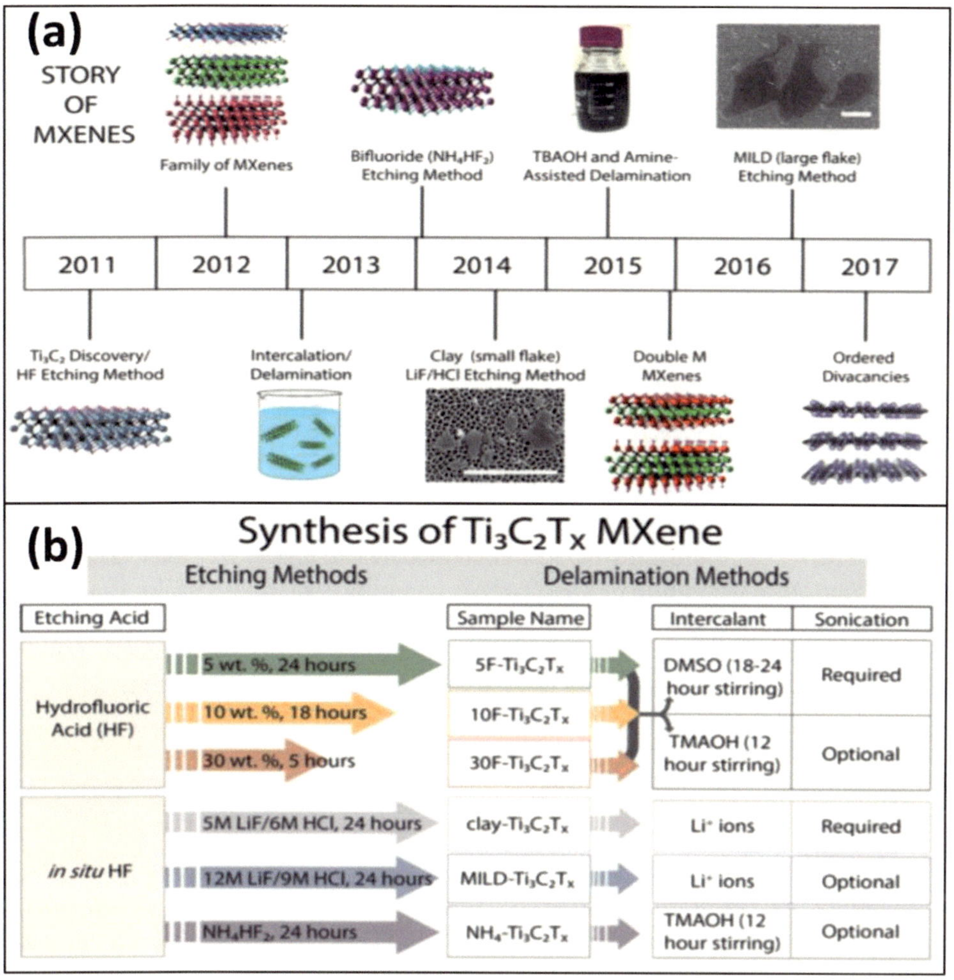

Figure 3.
(a) Timeline of synthesis of MXenes, (b) different protocols to synthesis different kind of MXenes (Source: Reprinted from [25] with permission from, copyright@2017, ACS).

delaminated MXene thick films by simply varying the etching and sonication conditions [35]. Although, Theoretical investigations shows the high electronic conductivity in MXenes. But there is still lack of knowledge and experimental expeditions to synthesize such exceptional MXene with control over surface chemistry.

2.2.2 Surface morphology

MXenes are synthesized by MAX precursors where M atoms are close packed and X atoms at interstitial sites [36]. Generally, MXenes have hexagonal close-packed structure with different order of M atoms in which M_2X follows ABABAB type order with hexagonal-stacked packing while M_3X_2 and M_4X_3 follows ABCABC type order with face-centered cubic stacking [22]. A study published by wang et al., Surface moieties play a key role in altering the properties of MXenes. The orientation and interaction between the terminal groups like -OH, -O, -F etc. strongly enhances interlayer hydrogen bonds which further improves the quality of MXenes. [37]. Also, the hydrogen bonding in MXene is highly influenced by surface terminations and interlayer spacing. Depending upon the occupancy of a functional group like -OH, -F and -O etc. the properties can be tuned for the respective application.

Intercalating the MXenes with ions, further gives a chance to mitigate the restacking behavior of MXene sheets for better performance, leading to display clay-like-behavior [38]. There is a lot to study and prepare pure MXenes for future energy applications.

2.2.3 Mechanical properties

MXenes exhibit peculiar physical and chemical properties which directly contribute to their mechanical behavior such as young's modulus, stiffness, defect generation, surface and elastic properties. Defect-rich MXenes with different terminal groups has strong covalent bonding with transition metal ion. Overall, there are various parameters which can be tunable to produce high performance MXenes. There are several theoretical studies on the mechanical, electronic as well as thermal properties of different types of MXenes [31, 39–43]. Experimentally the young's modulus of $Ti_3C_2O_2$ and Ti_3CO_2 was found to be 466 GPA and 983 GPA [44], these values were almost closer to the value predicted by theoretical simulations of 502 GPA [45]. Theoretical studies claims that M_2X exhibit much stronger in contrast to M_3X_2 and M_4X_3. But there is no experimental evidence to prove. However, in a study [46], A 5 μm thick paper film of $Ti_3C_2T_x$/PVA composite was able to hold~15,000 times its weight, which is evident to its strong wear and tear resistance property. Based on the surface terminations, there is a chance to modify surface properties of MXenes. Further investigations are needed to tune and enhance its nature.

3. Micro-fabrication techniques

3.1 Photolithography

Photolithography is a most promising technique at industrial scale which enables on-chip fabrication of high-resolution interdigital patterns of microelectrochemical systems (MEMS), Integrated Circuits (ICs), and complementary metal-oxide-semiconductors (CMOS) devices on various substrates with the help of computer-generated photomasks and photoresist designs. Recently, Jiang *et al*. fabricated all MXene based microsupercapacitors (MSCs) by spray coating on O_2 plasma treated predefined photoresists patterns prepared by direct photolithography on silicon substrates [47]. The thickness of the MXene coating was varied to get superior electrical conductivity. Device showed an ultra-high scan rate stability upto 300 V/s. Similarly, Kim and group fabricated a high performance MXene/CNT based hybrid on-chip MSCs by using Focused-ion beam (FIB) lithography technique [48]. They were able to fabricate the sub~500 nm gap between the fingers (W_g) reducing the active electrode foot print (W_e) of the device. Increasing the ratio, W_e/W_g by decreasing the gap between the electrode, which further shortens the ohmic losses. Hence improves the on-chip MXene based areal capacitance upto 317 $mFcm^{-2}$ at 50 mVs^{-1}. The capacitance retention of 32.8% was achieved even at higher scan rate of 100 Vs^{-1}. Due to imperfect resolution and ultra-narrow interelectrode gap. There are potential risks of degradation and a short circuit between the electrodes [49]. But, due to lack of electrode materials and stability issues, photo-lithography is not been used at its threshold. Xue and co-workers successfully demonstrated the electrophoretic deposition (EPD) of MXenes on the pre-patterned current collector in acetone solvent. EPD method is also grabbing attention in the scientific community [50]. There is a big room to do research. Since, photolithography technique comes out to be a challenging yet rewarding in terms of industrial fabrication and scalability for micro-supercapacitor applications.

3.2 Inkjet printing

Inkjet printing is a very popular technique for fabricating MSCs with excellent precision of designed pattern on various non conducting substrates. Nowadays, inkjet printing is gaining momentum in the scientific community. One can get desired thickness of printed layer to meet the certain applications. As name suggests inkjet printing is solely depends upon the prepared liquid ink.

There are three basic parameters which defines the behavior of the liquid inks:

1. Reynolds No. (R_e)

$$R_e = \frac{v \rho_a}{\eta}$$

2. Weber (W_e)

$$W_e = \frac{v^2 \rho_a}{\gamma}$$

3. Ohnesorge (O_h)

$$O_h = \frac{\sqrt{We}}{Re.}$$

Where ρ is the density (Kg/m^3), η is the dynamic viscosity ($N.S/m^2$), γ is the surface tension (N/m), v is the velocity (m/s), a is the nozzle diameter (m) [5]. To generate, stable ink droplets, numerical simulations have demonstrated that the rheology behavior of the ink should be in the range of $1 < Z < 10$ for better results. Also, to predict the rheological characteristics of a drop of ink, the inverse Ohnesorge number Z is used i.e., $Z = 1/O_h$. [51]. As increase in demand of self-charged and wearable devices,highly functionalized conductive$Ti_3C_2T_x$ (MXene) has attracted attention to directly prepare a highly stable conductive ink in various organic solvents. Inkjet printing is the cheapest and most viable technique to fabricate MXene based MSCs. Recently, Zhang et al. fabricated additive-free MXene ink based MSCs on flexible substrate shown in **Figure 4(a)**, N-Methyl-2-pyrrolidone (NMP) based MXene ink shows excellent volumetric capacitance of 562 F/cm^3 by inkjet printing. Demonstrating stable ink formulation in different organic solvents displayed in **Figure 4(b)**. The extrusion printed patterns exhibits power density as high as 11.4 μWcm^{-2}. Also, by adjusting the printing pass, the authors were able to reduce the sheet resistance upto 35 Ω/sq. from 445 Ω/sq. the areal capacitance and cycling stability of inkjet and extrusion printed MSC is given in **Figure 4(c,d)**. This study opens a new technique to fabricate low cost MXene ink-based MSC devices [52]. MXene aqueous ink with excellent oxidation resistance power were directly printed on paper substrate. The hybrid MXene suspension capped with sodium ascorbate (SA) displays the superior stability of upto 20 days. Due to its oxidation resistance nature and large interlayer spacing the conductivity of SA MXene improves to 119 Scm^{-1} this shows that there is still big room to develop MXene ink based printable devices for MSC Application [53].

3.3 Laser scribing

Laser engraving is another emerging cost-effective technique for the fabrication of MSCs on various customized substrates. Precise resolution with fast scanning speed makes this technique a superior approach in the field. But with all above benefits, there are few difficulties faced during the optimization of wavelength,

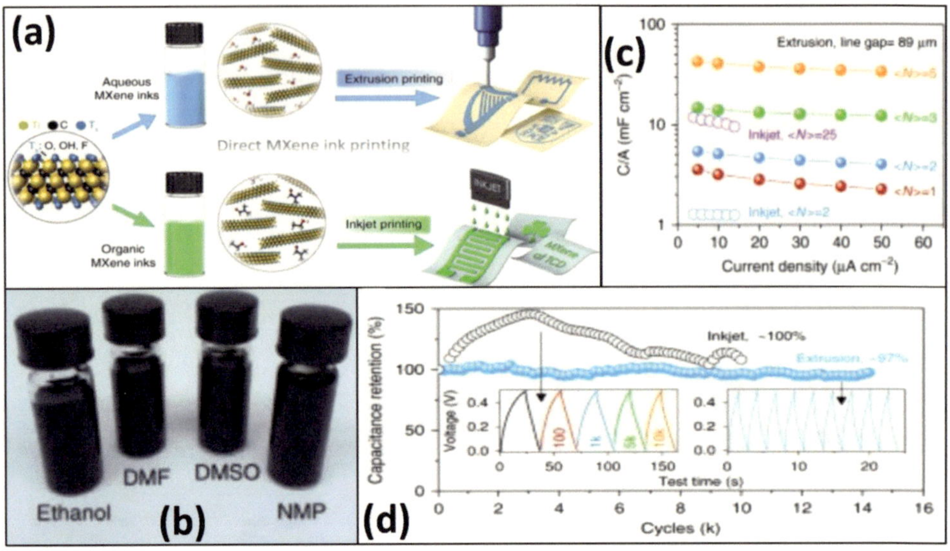

Figure 4.
(a) Schematic of direct printing of aqueous MXene inks used for extrusion printing and organic MXene inks used for inkjet printing on various substrates, (b) plot showing the areal capacitance of inkjet and extrusion printed MSC with various printing passes <N > = 2 and 25, (c) picture of different MXene organic inks, (d) cycling stability of inkjet and extrusion at current densities of 14 and 200μAcm^{-2}. (Source: Reprinted from [52], permission from, copyright @2019, Nature).

resolution and accurate speed suitable for the fabrication of MSCs on different substrates. Tang and co-group demonstrated the direct laser writing of $Ti_3C_2T_x$ interdigital electrodes by tuning the direction of laser scanning and rate. Interlayer spacing of restacked MXene was increased due to high photothermal oxidation effect of direct laser writing which enhanced the ion transport nature of the films [54]. Wang et al., fabricated double sided flexible asymmetric MSCs on thin nickel sheet by using spray coating technique followed by cutting of interdigital patterns by UV laser. By increasing the mass of active material, the maximum capacitance improved to 34 mF cm^{-2} approximately double as compared to the previous one. The fabricated double-sided device displayed considerable energy density of 2.62 μWhcm^{-2} at 2 mA cm^{-2} [55]. Further, Kurra and co-members reported a high areal capacitance based on clay-like MXene MSCs fabricated directly on paper by using a CO_2 laser. Clay-like MXene shows superior power density of 46.6 mWcm^{-2} at energy density of 0.77 μWhcm^{-2}, opening a new method to fabricate on-paper MSC devices [56].

3.4 Screen printing

Screen printing technology is one of the popular traditional technique to transfer predesigned ink patterns of active materials on various substrates. Important features of this technique are its scalability, reproducibility and repeatability. The screen printing technique is simple with high efficiency unlike other printing techniques, showing the enormous opportunity to explore. This inexpensive method can be used manually or even by automated machines. Generally, the setup includes the mesh screen with little gap between the substrate. With the help of squeegee, the ink is flooded over the screen mesh to print on the substrate [57]. Screen printing gained considerable attention to directly print MXene ink-based electrodes directly on a target substrate. Screen printing solely depends on rheological properties of ink which should be highly viscos and show a good shear thinning behavior. Additionally, the size and resolution of electrodes depends on mesh size. Recently, screen printed MSCs were fabricated on paper by using homogeneous ink of MXene sediments. The perfectly tuned thickness of MSCs reduced the sheet resistance upto 2.2 Ω sq^{-1} and gives excellent electrical conductivity of upto 450 Scm^{-1}. The energy density reached to 158 $mFcm^{-2}$ which is highest of its kind [58]. A two-step screen printing technique is employed to fabricate asymmetric MSC with interdigital pattern on paper as well as PET substrate. High energy density of 8.8μ$Whcm^{-2}$ in PVA-KOH was observed, far superior than many of other reports [59]. There are advantages of the technique to get high mechanical stability on different substrates which definitely enhances the electronic conductivity of the fabricated devices.

3.5 3D printing

3D printing technology has attracted lot of attention for scalable fabrication of 3D architectures for the development of small and portable electronics. Recently a new trend has been introduced to fabricate 3D MSCs. This method is less complex and easy to handle compared to other lithography techniques. MXenes are the emerging material to be introduced by this technique for the fabrication of gel-type ink-based 3D MSC device to enhance its areal and volumetric capacitance. Recently, Orangi et al., fabricated an ultra-stable gel-type MXene ink based MSC given in **Figure 5(a)** by modifying its viscoelastic behavior in universal water solvent. The as fabricated device **Figure 5(b)** displays a maximum energy density of 51.7μ$Whcm^{-2}$. Further optimizations of active material layer have been done to further enhance the areal capacitance. Good adhesion & no change of electrochemical performance, even on applying stress and strain on the device given in **Figure 5(c, f)**. The cyclic voltametric curves of all MSCFs and the best performing MSCF-10 as shown in **Figure 5(d, e)** [60]. Similarly, Free standing $Ti_3C_2T_x$ (MXene) ink-based 3D MSCs were fabricated followed by freeze drying for shape retention. To increase the

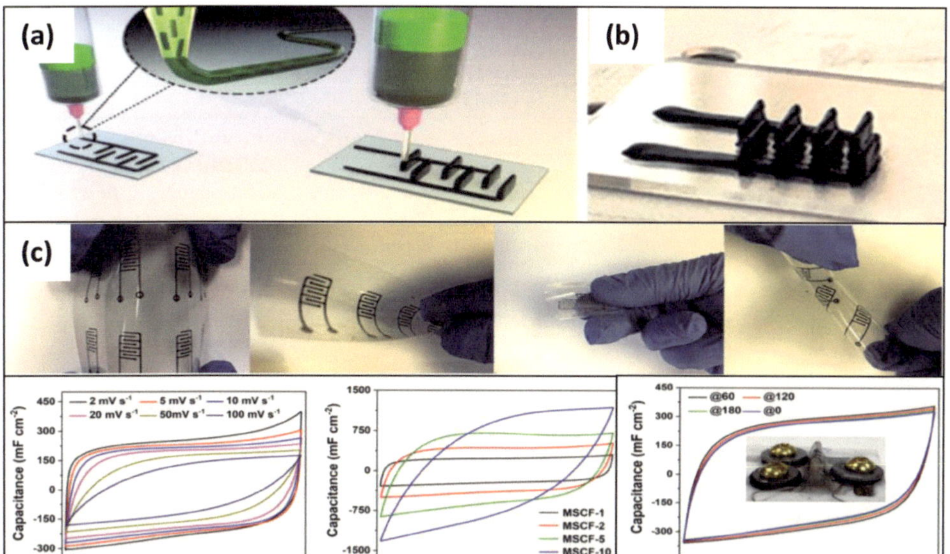

Figure 5.
(a) Demonstration of 3D interdigital MSC by 3D printing technique, (b) As printed MSC on glass substrate, (c) picture showing the strong adhesion of printed pattern on plastic substrate with great flexibility during repeated bending cycles, (d) CV curves of MSCF-1 at different scan rates from 2mVs^{-1} to 100 mVs^{-1}, (e) CV curves of various MSCFs at scan rate of 5 mVs^{-1}, (f) CV curves of MSCF-10 at scan rate of 10 mVs^{-1} at different bending angles (Source: Reprinted from [60] with permission from, copyright@2020, ACS).

stability and electrical conductivity, optimizing the mass loading to get the better viscoelastic behavior are the key parameters to obtain high areal capacitance. The maximum areal capacitance of 2.1 Fcm^{-2} at 1.7 $mAcm^{-2}$ was achieved by a single MSC device. This unique technique has a wide base to explore micro-supercapacitor applications just by playing with the rheological properties of inks [61].

3.6 Other techniques

Unconventional methods have also been employed to fabricate the MXene based MSCs. A group reported the direct writing of highly concentrated MXene-in-water inks of upto 30 mg/mL in water on different substrates by using commercial roller ball pen. Interdigital electrodes were designed to fabricate Micro-supercapacitors. Areal capacitance of single MXene MSC was 5 $mFcm^{-2}$ and by joining four MSC devices in series, the potential window reached upto 2.4 V which is evident for the development of flexible MSC devices [62]. Zhang *et al.* used a novel stamping technique to fabricate interdigital MSC on various substrates by using Ti_3CNT_x (MXene) inks. They observed an areal capacitance upto 61 $mFcm^{-2}$ at 25 μAcm^{-2} which outperforms many of previous reports. The device also exhibits high coulombic efficiency of 100% even after 10,000 cycles. This novel approach opens a new exciting method to fabricate MSC in easy and facile way [63]. Hue *et al.* demonstrated a facile two-step laser jet vacuum assisted filtration approach to fabricate all-solid-state MXene based symmetric microsupercapacitors followed by gold sputtering on regular A4 paper. The device exhibits high energy density in the range of 5.48–6.1 $mWhcm^{-2}$ depending upon the deposited thickness of the electrode. The maximum areal capacitance of 27.29 $mFcm^{-2}$ was achieved. This

Techniques	Method	Merits	Demerits
Photolithography	Direct	Wafer-scale manufacturing, uniform & high-resolution patterning [47, 48].	Multi-step process, template assisted, time consuming method [48].
Inkjet and Extrusion Printing	Indirect	Scalable production, customized design, less wastage of material [52, 62].	Uncontrollable procedure of ink synthesis, Low resolution, nozzle jamming is one of the main disadvantages of this technique [52].
Laser Scribing	Direct	Cost effective, fast simple, high controllability [54, 55].	Confined to very few types of materials [6].
Screen Printing	Indirect	Highly scalable and fast process [59].	Relatively low-resolution power
3D Printing	Indirect	Controllable design of patterns, versatile thickness control [61].	Limited to few materials, complex processibility [61].
Electrophoretic Deposition	Direct	Economically viable, facile procedure [50].	Limited applicability.
Vacuum-assisted-filtration	Indirect	Easy process, controlled thickness [67].	Low resolution, size and shape limited.

Table 1.
Merits and demerits of various fabrication techniques of MSCs.

simple strategy of laser jet printed mask-assisted technique exhibits the potential for low cost fabrication method without compromising with device performance [64]. Li and co-workers proposed a simple scratch method to fabricate $Ti_3C_2T_x$/EG (MXene/exfoliated graphene) based MSC. A common syringe was employed with custom made X/Y axis instrument to fabricate the interdigital patterns. The device was able to display electrochemical stability upto 5000 charge/discharge cycles with around 90% retention of capacitance. This new approach shows promising results with almost negligible cost of fabrication at large scale [65]. Similarly, another group used automated scalpel technique to carve semi-transparent PEDOT/$Ti_3C_2T_x$ heterostructures micro-supercapacitors. Device exhibit considerably high capacitance of 2.4 $mFcm^{-2}$ at 10 mVs^{-1} shown by 100 nm device with almost 58% of capacitance retention at scan rate of 1000 mV/s. Further changes in color were observed on applying voltage 0.6 to 0 V and – 0.6 to 0 V while discharging which displays good electrochromic behavior PEDOT/$Ti_3C_2T_x$ MSCs [66]. Advantages and disadvantages of various fabrication techniques of MSCs can be seen in **Table 1**.

4. MXene and its 2D hybrids for micro-supercapacitors

4.1 MXene based materials

In the past few years, MXenes have shown promising results for micro-supercapacitor applications. Due to their unique morphology, high metallic conductivity~10,000 Scm^{-1} and excellent intercalation behavior. Kurra *et al.* reported all MXene based low cost and highly scalable coplanar microsupercapacitors on paper substrates, the clay like MXenes based MSC displays the electrical conductivity of 128 Scm^{-1} and areal capacitance of 25 $mFcm^{-2}$ in PVA-H_2 SO_4 gel electrolyte. This study suggests the thickness of the active material plays a key role in the enhancement of the areal capacitance [56]. Similarly, Jiang and co-workers reported a wafer scale approach to fabricate an on-chip MXene based MSC device. The typical procedure includes photolithography of interdigital patterns followed by spray coating. The optimized $Ti_3C_2T_x$ – 0.3 μm exhibits more capacitive behavior. The fabricated device was able to convert constant output positive peak voltage of 0.6 V into~0.56 V which is comparable with commercially available capacitor (4mF). Demonstrating the advancements of MXene based MSCs for better alternative than bulky electrolytic capacitors in circuits [47]. Peng *et al.* fabricated interdigital patterned device by spray coating of $Ti_3C_2T_x$ flakes directly on glass substrates which shows considerable areal capacitance of 19.6 $mFcm^{-2}$ at 20 mVs^{-1} with ultra-high volumetric capacitance of 356.8 Fcm^{-3} at 0.2 $mAcm^{-2}$ which is better than many of the carbon materials reported in the literature. But, the significant increase of areal capacitance 27.3 $mFcm^{-2}$ at 20 mVs^{-1} can be seen by introducing the platinum current collectors [68]. Recently, A new strategy has been employed to pattern semi-transparent film of MXene based hybrid device on glass substrate without using any mask. An automated scalpel tool was used to produce micropatterns at various levels of transparency. On the increase of transparency from 38–88%, areal capacitance from 19 to 283 μFcm^{-2} can be evidently seen to be increased because of thick layers of MXenes. In contrast, With the increase of coating cycle, the resistance also increases from 0.8 to 2 kΩ. The device demonstrated excellent capacitive behavior, offers variety of tunable approach by which one can enhance its physiochemical properties [69]. Li et al., reported the fabrication of a double sided MSCs (DSMSCs) based on MXene ink with high working potential window of 7.2 V connected in different series and parallel configurations. With the decrease of interspace between MXene electrodes, the steep rise of capacitance can be seen.

Hence, DSMSC with 10 μm interelectrode gap displays the highest volumetric capacitance of 308 Fcm^{-3} at 5 mVs^{-1} with ultra-high coulombic efficiency of 96.4% even after 10,000 cycles [70]. Quain and group reported direct writing with pen using additive-free MXene ink on flexible paper and non-paper substrates. The ink suspension displays good polydispersity index of 0.549 which consists of both small and large flakes of MXene at 30 mg mL^{-1} This single step fabrication technique is used to write on various flexible substrates. High potential window upto 2.4 V was also achieved by connecting four MSCs in series [62]. A facile Freeze-and -Thaw-assisted method (FAT) was used to produce two-atom thin layers of MXene with extra ordinary strength and flexibility. FAT-MXene exhibits an Areal capacitance of 23.6 $mFcm^{-2}$ with high volumetric capacitance. 591 Fcm-3 at 20 mVs^{-1} [71]. Zhang and group-fabricated a flexible asymmetric microsupercapacitors comprising MXene and MXene-MoO_2 films as negative and positive electrodes. The fabrication process includes vacuum filtration of the films followed by laser cutting of the interdigital patterns as given in the schematic **Figure 6(a)**. The asymmetric device exhibits large potential window of 1.2 V which is almost double of symmetric device. The device delivers volumetric capacitance of 63.3 Fcm^{-3} and the CV and GCD curves shown in **Figure 6(b, c)** with excellent capacitance retention of 88% after 10,000 cycles **Figure 6(d)** [72]. Huang *et al.* reported a facile strategy to produce free standing-thick MXene sheets by vacuum filtration. The films exhibit an ultra-high conductivity upto 1.25×10^5 Sm^{-1} for flexible-MSC. Further efforts has been done to fabricate an interdigital patterned MSC device which displays an considerable areal capacitance of 340 $mFcm^{-2}$ with volumetric capacitance of 183 Fcm^{-3} and the corresponding energy density and power density are 12.4 $mWhcm^{-3}$ and 218 $mWcm^{-3}$ [73]. Another group demonstrated a highly conductive paper based MXene electrodes possessing a reasonable areal capacitance of 23.4 $mFcm^{-2}$ at 0.05 $mAcm^{-2}$. One step process fabrication of electrodes in series as well as parallel to further get the desired capacitance [74].

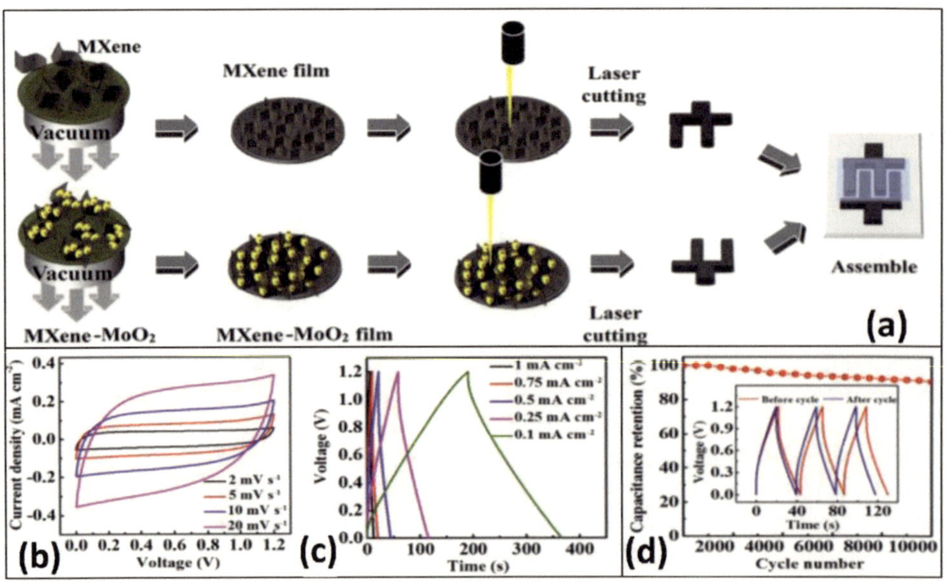

Figure 6.
(a) Fabrication procedure of MXene//MXene-MoO_2-AMSCs asymmetric MSC, (b) cyclic voltametric curves at different scan rates from 2 to 20 mVs^{-1}, (c) Galvanostatic charge–discharge at different current densities from 0.1 to 1mAcm^{-2}, (d) cyclic stability at current density of 0.5 $mAcm^{-2}$. Inset shows the charge–discharge cycles after and before10,000 cycles (source: Reprinted from [72] permission with Elsevier).

4.2 MXene and carbon materials

Recently, Kim et al., reported a scalable production of MXene/CNT based MSCs with a 500 nm gap between the interdigital fingers exhibiting fast ion diffusion for superior conductivity. High areal capacitance of 317.3 $mFcm^{-2}$ was achieved at 50 mVs^{-1} by composite of S-DWCNT/MXene in PVA-H_2SO_4 gel electrolyte. It is also observed that by decreasing the electrodes gap 10 μm to 500 nm, improves the ionic transfer rate, leading to increase in areal capacitance and energy density [48]. A 3D MXene/rGO self-healable aerogel MSC were reported by Yue and group. The fabrication process is shown in **Figure 7(a, b)** They employed new approach to fabricate highly stable device by keeping in focus to real time applications. Fabricated device was encapsulated into self-healing Polyurethane (PU) which enabled the device to adhere the external damage. The composite aerogel exhibited an exceptional recovery of electronic and mechanical properties even

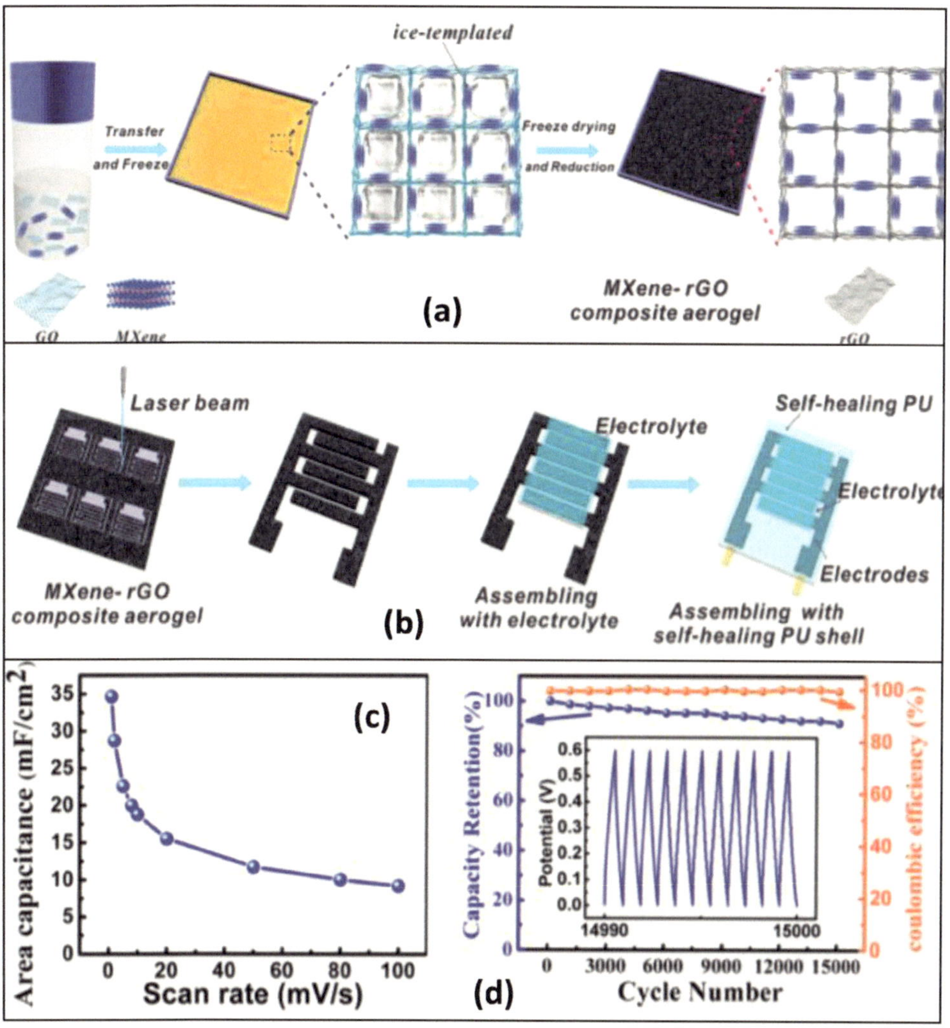

Figure 7.
(a) Fabrication procedure of MXene-rGO composite aerogels, (b) laser cutting of interdigital pattern on MXene-rGO composite followed by assembling with self-healing PU, (c) graph showing the areal capacitance vs. scan rate MXene-rGO composite, (d) cycling stability of MXene-rGO composite aerogel MSC at 2 $mAcm^{-2}$ (inset showing the GCD curves from 14990th to 15000th cycles (source: Reprinted from [75] with permission from, copyright@2018, ACS).

after full breakdown and shows the areal capacitance of 34.6 mF cm^{-2} at 1 mVs^{-1}, the areal capacitance and Cycling stability is shown in **Figure 7(c, d)** [75]. Couly *et al.* fabricated a high performing asymmetric flexible micro-supercapacitor based on MXene as negative and rGO as positive in both sandwich as well as interdigital configurations by using simple spray-coating of active material on PET substrates. The working potential window increased to 1 V for asymmetric device even with no. of bending and folding cycles, the maximum areal capacitance of 2.4 $mFcm^{-2}$ at 2 mV/s was achieved. This study shows MXene as a promising material for negative electrode in asymmetric configuration with good stability and robust performance [76]. There is still wide room for further exploitation of carbon-based materials for micro-supercapacitor applications. A new emerging trend to produce nanofibers based on yarn type super capacitors for self-charged and wearable energy storage devices. MXenes have shown great potential to produce textile-based energy storage devices due to its robust stability as well as extraordinarily tunable properties. Yu and group, reported a helical shape MXene/CNT scaffold hybrid structure with reasonable volumetric capacitance of 19.1 Fcm^{-3} at 1.0 Acm^{-3} in 6 M of aqueous LiCl electrolyte. The MXene/CNT fiber exhibit good Energy density of 2.55 to 1.15 $mWhcm^{-3}$ at power density of 0.046 to 1.82 W cm^{-3} in LiCl gel electrolyte. The best performing device displays the capacitance retention of 19.5 Fcm^{-3} (84%) at current density of 1.0 Acm^{-3} [77]. MXene/rGO hybrid fiber supercapacitors were fabricated by wet-spinning assembly strategy with extremely high volumetric capacitance of 586.4 Fcm^{-3} at 10 mV/s. The composite fibers display an ultra-high electrical conductivity of 2.9×10^4 S cm^{-1}. They observed that the flexibility of the fiber can be increased by adjusting the concentration of graphene [78].

In another report by chen and group, MXene-MoS_2 based free standing MSCs were fabricated by simple and low-cost vacuum filtration method followed by carving of interdigital patterns with laser source. By introducing the MoS_2 into MXenes which further enhances the electrochemical performance with almost 60% increase as compared to pristine MXene. i.e., the fabricated device displays a high specific capacitance of 173.6 F/cm^3 at the scan rate of 1 mV/s, MSC shows around 98% of capacitance retention with 89% of coulombic efficiency even after 6000 cycles along with several bending angle of device upto 150°. The above study demonstrated huge potential of TMDs which can be introduced with MXenes to make high performing MSC devices [67]. Li *et al.* demonstrated a strategy to mitigate the self-restacking of MXene layers by introducing RuO_2 nanoparticles by simple wet chemical phase reaction to improve the ion exchange rate. Also integrating with conductive Ag nanowires into the MXene further decrease the surface resistance of electrodes. The optimized MSC device achieved an ultrahigh volumetric capacitance of 864.2 Fcm^{-2} at 1 mV/s with 90% of capacitance retention even after 10,000 cycles [79]. For the first time, Wang *et al.* reported PANI/MXene based film electrodes with an exceptionally high volumetric capacitance of 1167 Fcm^{-3}. The asymmetric device by taking MXene as a negative electrode exhibit a maximum energy density of 65.6 WhL^{-1} which overshadows many of the previous reported MXene based Micro-supercapacitors [80]. A new kind of stretchable micro-supercapacitors based on MXene/Bacterial Cellulose (MXene/BC) composite free-standing paper were fabricated showing an exceptionally high young's modulus of 15–35 GPa with tensile strength of upto 200–300 GPa. Here BC acts as a spacer intercalated between the MXene sheets to prevent the re-stacking of MXene flakes. A conventional laser cutting tool used to fabricate stretchable micro-supercapacitor device was prepared which displays the high areal capacitance of 111.5 mF cm^{-2} in parallel device configuration with reasonable energy density of 0.00552 $mWhcm^{-2}$ [81]. Shao *et al.* synthesized MXene-polymer composite nanofibers as flexible yarn electrodes by simple electrospinning the active material on PET sheets. The symmetric device displays high areal capacitance of upto 18.39 $mFcm^{-2}$ at scan rate of 50 mVs^{-1} which is better than many other carbon based yarn fiber supercapacitors [82]. Another group of researchers fabricated MXene/ PEDOT-PSS based yarn supercapacitors (YSCs). A 3 cm flexible fiber shows extraordinarily high length capacitance of 131.7 mF cm^{-1} at 0.2 $mAcm^{-1}$ with capacitive retention of 95% even after 10,000 cycles. They observed the reasonable contribution of conductive-polymer PEDOT-PSS in improving the device performance, suggesting a potential candidate in flexible yarn supercapacitor in portable electronics [83]. A new strategy has been employed to fabricate dual-core yarn supercapacitor (YSC), fabrication process shown in **Figure 8(a, b)** consist of rGO and MXene hybrid fibers encapsulated with PVA-H_2SO_4. The average diameter of YSC is~500μm showing the superior linear capacitance 43.6 $mFcm^{-1}$ at 20 mVs^{-1}. The areal capacitance was maintained above 175 $mFcm^{-2}$ with respect to increasing length. They observed the charge transfer resistance (Rct) ESR of YSCs decreases gradually with increase in length such as~30.3 Ωcm^{-1} at 3 cm, 3.9 Ωcm^{-1} at 10 cm to 1.6 Ωcm^{-1} at 15 cm the graphs are shown in **Figure 8(c, d)**. The YSC device of 15 cm displayed areal density of 54.5 $\mu Whcm^{-2}$ at a power density of

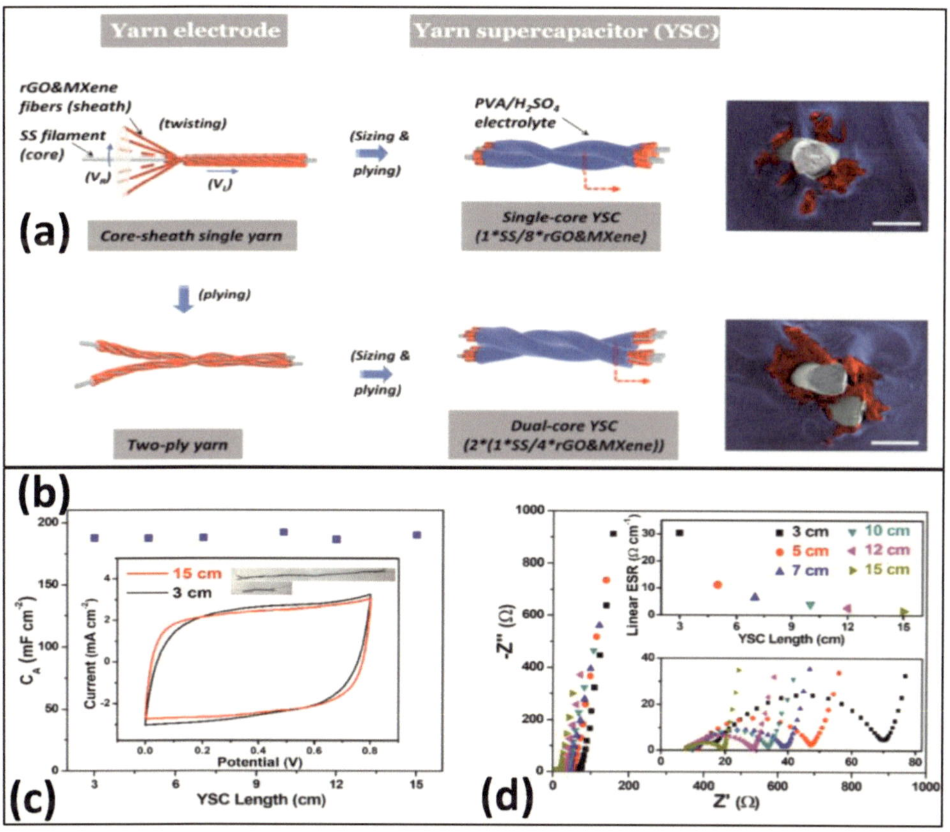

Figure 8.
*Schematic illustrations of fabrication process of (a) 1*SS/S* rGO&MXene YSC single core YSC cross-sectional SEM image (scale bar -100 μm), (b) (2*(1*SS/4*rGO&MXene)) dual-core YSC cross-sectional SEM image (scale bar -100 μm), (c) areal capacitance of dual-core YSC inset: CV curves at 20mVs^{-1} in 3 and 15 cm, (d) comparison of Nyquist plots of dual-core YSCs from 1 MHz to 0.01 Hz in different length inset: Linear ESR as a function of YSC length (top), zoom in image of Myquist plots (bottom) (source: Reprinted from [84]with permission from, copyright@2020, ACS).*

Material	Method	Electrolyte	Potential Window	Device Performance		Specific Capacitance		Capacitance Retention	References
				Energy Density	Power Density	Areal	Volumetric		
$Ti_3C_2T_x$ (100 nm-25 μm)	Photo-lithography		0 to 0.6	—	PVA-H_3PO_4	0.5 $mFcm^{-2}$ @120 Hz	30 Fcm^{-3} @120 Hz	—	[47]
$Ti_3C_2T_x$/CNT $_{500nm}$	FIB Lithography	PVA-H_2SO_4	0 to 0.6	—	—	317 $mFcm^{-2}$ @ 50mVs^{-1}	—	—	[48]
$Ti_3C_2T_{x\,N=25}$ $Ti_3C_2T_{x\,N=5}$	Inkjet Extrusion	PVA-H_2SO_4	0 to 0.5	— 0.32 $\mu Whcm^{-2}$	— 11.4 μWcm^{-2}	12 $mFcm^{-2}$ —	562 Fcm^{-3} —	100% (10,000) 97% (15,000)	[52]
SA- $Ti_3C_2T_x$ P-$Ti_3C_2T_x$	Inkjet Inkjet	PVA-H_2SO_4	0 to 1	100.2 $mWhcm^{-3}$ —	1.9 Wcm^{-3} —	108.1 mF cm^{-2} @1 Ag^{-1} 48.4 $mFcm^{-2}$ at 1Ag^{-1}	720.7 Fcm^{-3} @1 Ag^{-1} —	94.7% (4,000) 72.4% (4,000)	[53]
$Ti_3C_2T_x$	Laser Writing	3 M H_2SO_4	0 to 0.6	0.25 $\mu Whcm^{-2}$	2.94 $mWcm^{-2}$	15.03 $mFcm^{-2}$	—	105% (10,000)	[54]
Double sided Zn//MXene (Asymmetric) Carbon//MXene (Asymmetric)	Laser writing Laser writing	PVA- Zn $(CF_3SO_3)_2$ PVA-LiCl	0 to 1.1 0 to 0.8	— 2.62 $\mu Whcm^{-2}$	— —	66.5 $mFcm^{-2}$ 52.3 $mFcm^{-2}$ @2$mAcm^{-2}$	— —	86% (5,000) —	[55]
Clay like $Ti_3C_2T_x$	Laser Writing	PVA-H_2SO_4	0 to 0.6	0.77 $\mu Whcm^{-2}$	46.6 $mWcm^{-2}$	25 $mFcm^{-2}$	—	92% (10,000)	[56]
$Ti_3C_2T_x$ Sediments	Screen Printing	PVA-H_2SO_4	0 to 0.6	1.32 $\mu Whcm^{-2}$	778.33 μWcm^{-2}	158 $mFcm^{-2}$	—	95.8% (16,000)	[58]
MXene//Co-Al layered double hydroxide (Asymmetric) MXene	Screen Printing Screen Printing	PVA-KOH PVA-KOH	0.4 to 1.45 0 to 0.6	8.84 $\mu Whcm^{-2}$ 3.38 $\mu Whcm^{-2}$	0.23 $mWcm^{-2}$ —	40.0 mF cm^{-2} @0.75b$mAcm^{-2}$ 25 $mFcm^{-2}$	— —	92% (10,000) —	[59]
$Ti_3C_2T_x$	3D Printing	PVA-H_2SO_4	0 to 0.6	8.4 $\mu Whcm^{-2}$	3.7 $mWcm^{-2}$	168.1 $mFcm^{-2}$	—	—	[60]

Material	Method	Electrolyte	Potential Window	Device Performance		Specific Capacitance		Capacitance Retention	References
				Energy Density	Power Density	Areal	Volumetric		
$Ti_3C_2T_x$	3D Printing	PVA-H_2SO_4	0 to 0.6	0.0244 $mWhcm^{-2}$	0.64 $mWcm^{-2}$ @ 4.3 $mAcm^{-2}$	2.1 Fcm^{-2} @1.7 $mAcm^{-2}$	———	90% (10,000)	[61]
$Ti_3C_2T_x$	Direct Writing	PVA-H_2SO_4	0 to 0.6	———	———	5$mFcm^{-2}$	———	———	[62]
l-$Ti_3C_2T_x$	Stamping Strategy	PVA-H_2SO_4	0 to 0.6	0.63 $\mu Whcm^{-2}$	0.33 $mWcm^{-2}$	56.8 $mFcm^{-2}$ @ 10mVs^{-1}	———	93.7% (10,000)	[63]
$Ti_3C_2T_x$	Laser jet Printing	PVA-H_2SO_4	0 to 0.6	6.1 $mWhcm^{-3}$	———	27.29 $mFcm^{-2}$ @0.25 $mAcm^{-2}$	———	———	[64]
$Ti_3C_2T_x$	Scratch method	PVA-H_3PO_4	0 to 0.7	2.3 $mWhcm^{-3}$	159.6 $mWcm^{-3}$	25.5 $mFcm^{-2}$ @ 5mVs^{-1}	———	90% (5,000)	[65]
PEDOT/$Ti_3C_2T_X$ 100nm	Spray Coating	PVA-H_2SO_4	0 to 0.6	———	———	2.4 $mFcm^{-2}$ @ 10mVs^{-1}	———	———	[66]
Free-standing $Ti_3C_2T_x-MoS_2$	Laser Engraving	Gelatin-$ZnSO_4$	0 to 0.8	15.5 $mWhcm^{-3}$	0.97 Wcm^{-3}	———	173.6 Fcm^{-3} @1mVs^{-1}	98% (6,000)	[67]
s-Ti_3C_2Tx	Spray coating + Laser engraving	PVA-H_2SO_4	0 to 0.6	11–18 $mWhcm^{-3}$	0.7–15 $W cm^{-3}$	27.3 $mFcm^{-2}$ @ 20mVs^{-1}	356.8 Fcm^{-3} @ 0.2 $mAcm^{-2}$	100% (10,000)	[68]
90 nm $Ti_3C_2T_x$ thin film	Dip Coating + Automated Scalpel patterning	PVA-H_3PO_4	0 to 0.6	———	———	———	1500 Fcm^{-3}	———	[69]
$Ti_3C_2T_X$-MSC 10 μm	Laser Etched	PVA-H_2SO_4	0 to 0.6	———	———	———	308 Fcm^{-3} @5mVs^{-1}	93% (10,000)	[70]
$Ti_3C_2T_X$	Mask-assisted vacuum filtration	PVA-H_2SO_4	0 to 0.6	10.3 to 29.6 $mWhcm^{-3}$	18.6 to 3.1 Wcm^{-3}	23.6 $mFcm^{-2}$	591 Fcm^{-3}	97.8% (2,000)	[71]

Material	Method	Electrolyte	Potential Window	Device Performance		Specific Capacitance		Capacitance Retention	References
				Energy Density	Power Density	Areal	Volumetric		
$Ti_3C_2T_X$ // $Ti_3C_2T_X$-MoO_2-AMSCs (Asymmetric)	Vacuum filtration + Laser cutting	PVA- LiCl	0 to 1.2	9.7 $mWhcm^{-3}$	0.198 Wcm^{-3}	~19 $mFcm^{-2}$	63 Fcm^{-3} @ 2mVs^{-1}	88% (10,000)	[72]
$Ti_3C_2T_X$	Vacuum filtration + Laser cutting	PVA-H_2SO_4	0 to 0.7	43.5 $mWhcm^{-2}$ 12.4 $mWhcm^{-3}$	87.5mWcm^{-2} 218.8 $mWcm^{-3}$	73–340 $mFcm^{-2}$	183–162 Fcm^{-3}	82.5% (5,000)	[73]
$Ti_3C_2T_X$ on paper	Spray coating + Laser coating	PVA-H_2SO_4	0 to 0.6	1.48 $mWhcm^{-3}$	189.9 $mWcm^{-3}$	23.4 $mFcm^{-2}$ @0.05 $mAcm^{-2}$	———	92.4% (5,000)	[74]
$Ti_3C_2T_X$ -Graphene aerogel	Laser cutting	PVA-H_2SO_4	0 to 0.6	———	———	34.6 $mFcm^{-2}$ @ 1 mVs^{-1}	———	91% (15,000)	[75]
$Ti_3C_2T_X$//rGO (Asymmetric)	Spray coating	PVA- H_2SO_4	0 to 1	8.6 $mWhcm^{-3}$	0.2 Wcm^{-3}	2.4 $mFcm^{-2}$ @2 mVs^{-1}	80 Fcm^{-3}	97% (10,000) - Interdigital	[76]
$Ti_3C_2T_X$/CNT (YSC)	———	PVA-LiCl	0 to 0.9	2.55m$Whcm^{-3}$	45.9 $mWcm^{-3}$	———	22.7 Fcm^{-3} @ 0.1 Acm^{-3}	99% (1,600)	[77]
$Ti_3C_2T_X$/rGO	———	PVA-H_3PO_4	0 to 0.8	13.03 $mWhcm^{-3}$	0.59 Wcm^{-3}	———	586.4 Fcm^{-3} @ 10 mVs^{-1}	———	[78]
RuO_2/$Ti_3C_2T_X$	Screen printing	PVA-KOH	0 to 0.6	13.5 $mWcm^{-3}$	48.5 Wcm^{-3}	———	864.2 Fcm^{-3} @ 1mVs^{-1}	90% (10,000)	[79]
PANI/MXene// MXene	———	1 M H_2SO_4	0 to 1.4	65.6 WhL^{-1}	1687.3 WL^{-1}	———	231.4 Fcm^{-3} @ 10mVs^{-1}	87.5% (5,000)	[80]
MXene/Bacterial Cellulose	Vacuum filtration + Laser cutting	PVA-H_2SO_4	0 to 0.6	0.0055 $mWhcm^{-2}$	———	112.2 $mFcm^{-2}$	———	———	[81]
Polyester @MXene	Electrospinning of fibers	PVA-H_2SO_4	0 to 0.6	0.38–0.67 $\mu Whcm^{-2}$	0.09–0.39 $mWcm^{-2}$	7.99 $mFcm^{-2}$ – 18.39 $mFcm^{-2}$	~4.5 Fcm^{-3} @5 mVs^{-1}	98.2% (6,000)	[82]

DOI: http://dx.doi.org/10.5772/intechopen.97650

Material	Method	Electrolyte	Potential Window	Device Performance		Specific Capacitance		Capacitance Retention	References
				Energy Density	Power Density	Areal	Volumetric		
MXene/ PEDOT-PSS	Fiber coating	Conductive binder PEDOT-PSS	0 to 0.5	——	——	131.7 $mFcm^{-1}$ @0.2 $mAcm^{-1}$	——	90% (10,000)	[83]
rGO/MXene Hybrid	Wet-spinning	PVA-H_2SO_4	0 to 0.8	5.5 $\mu Whcm^{-1}$	510.9 μWcm^{-1} 2502.6 μWcm^{-2}	77 $mFcm^{-1}$ 377.3 $mFcm^{-2}$	23.2 Fcm^{-3}	82% (10,000)	[84]

Table 2.
Summary of recently reported MXene based micro-supercapacitors.

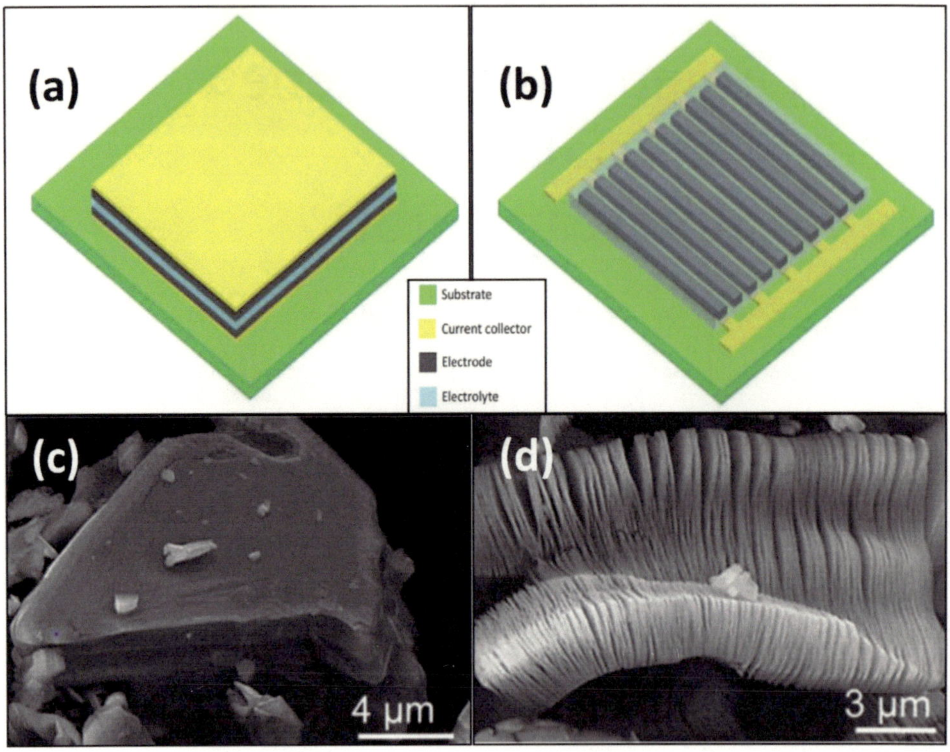

Figure 1.
Architecture of micro-supercapacitors: (a) sandwich type micro-supercapacitor, (b) interdigital patterned micro-supercapacitor. [2] (c) SEM image of Ti_3AlC_2 (MAX) phase, (d) SEM image of etched $Ti_3C_2T_x$ (MXene) phase [3].

Two-dimensional (2D) materials like graphene, h-BN, Transition metal dichalcogenides (TMDCs), black phosphorus (BP), MXenes and 2D metal oxides and hydroxides etc. are most widely used in energy storage applications due to their outstanding electronic, mechanical, optical and physio-chemical properties [6]. Carbon based materials including Carbon [7], carbide derived carbon [8], onion-like carbon [9], graphene [10], Carbon nano tubes (CNT) [11], laser scribed graphene [12] displays high electronic conductivity and relatively large surface area but due to electric double layer formation, they lack high energy density. Similarly, pseudocapacitive materials such as transition metal oxides like MnO_2 [13], MoO_3 [14], conductive polymers [15] as well as TMDs [16] which suffer from low electronic conductivity with reasonable power and cycling performance has been already explored in Micro-supercapacitor devices applications. But, MXenes have garnered great attention from the scientific community all over the world since their discovery in 2011, by Naguib and group [17]. A large family of two-dimensional early carbides, nitrides and carbon nitrides produced by selective etching of A layer (typically Al and Ga) from the precursor layered ternary carbides/nitrides (MAX phases). Their general formula is $M_{n+1}X_nT_X$ (n = 1, 2, 3), where M represents a Transition metal, X is carbon and/or nitrogen, T stands for surface termination groups (-F, -OH, -O etc.) [18]. In particular, the dual nature of MXenes that is superior ion transport due to inner transition metal carbide layer as well as property to exhibit fast redox reaction because of large active sites [18, 19]. MXenes combine high electronic conductivity of MAX phases as well as hydrophilic nature due to the surface terminations such properties make them a considerable candidate for a host of applications. $Ti_3C_2T_X$ is one of the most studied member of

1251.5 μWcm^{-2} which directly outperforms the previous reported literatures [84]. The detailed summary of data is presented in the **Table 2**.

5. Future perspective and outlook

Since the discovery of MXenes in 2011 by Naguib et al. [17] MXenes have become a best choice for micro-electrodes to develop on-chip and self-charged MSC for wireless and wearable electronics applications. There is a significant increase in research on MXene based MSC due to its extraordinarily high electronic conductivity, good volumetric capacitance and excellent advancement in properties.

But, the development of MXene based MSC are still in early stage with necessity of optimization of electrode material, suitable electrolyte, substrates and many more. Right now, the focal point of researchers is on the enhancement of areal capacitance and power density of the fabricated MXene based MSC devices. However, there is an act of negligence over its property to self-discharged in open-circuit which needs to be resolved as soon as possible. One solution to this is to further integrate MSCs device with energy harvester like micro-piezoelectric or solar power cell component which will improve long term charge-storage property instead of self-discharging.

Also, the choice of electrolyte plays an important role to enhance the electrochemical performance of MSC device. Generally, polymer gel electrolyte. Particularly, PVA-H_2SO_4 is widely used ion exchange for MXene based electrodes for micro-devices. But due to low voltage window there is a call for an alternative which can help to increase the stability and voltage window. So that, there is an urgent requirement to study different electrolytes and polymers to achieve better performing MSC. In contrast to polymer matrix electrolyte, a new emerging class of quasi-solid electrolyte called as ionogel which is more mechanically and thermally stable than the regular gel electrolyte. All this demonstrates the possibility of ionogel to be a potential candidate for MSCs. To further expand potential window there is a requirement to make asymmetric devices which can further increase the voltage range above 3 V for real time applications.

Despite recent developments of $Ti_3C_2Ti_x$ (MXene) based MSCs. There is still a big room to synthesis new MXene materials and explore their properties for the better understanding of charge storage mechanism which later can pave the way for future MSCs devices.

Acknowledgements

This work was financially supported by the Department of Science and Technology (DST)-SERB Early Career Research project (Grant No. ECR/2017/001850), DST-Nanomission (DST/NM/NT/2019/205(G), DST/TDT/SHRI-34/2018), Karnataka Science and Technology Promotion Society (KSTePS/VGST-RGS-F/2018-19/GRD NO. 829/315).

Conflict of interest

The authors declare no conflict of interest.

Author details

Aditya Sharma and Chandra Sekhar Rout*
Centre for Nano and Material Sciences, Jain University, Bangalore, India

*Address all correspondence to: r.chandrasekhar@jainuniversity.ac.in

References

[1] Wang J, Li F, Zhu F, Schmidt OG. Recent Progress in Micro-Supercapacitor Design, Integration, and Functionalization. Small Methods. 2019;3(8):1800367.

[2] Beidaghi M, Gogotsi Y. Capacitive energy storage in micro-scale devices: recent advances in design and fabrication of micro-supercapacitors. Energy Environ Sci. 2014 Feb 20;7(3):867-84.

[3] Michael J, Qifeng Z, Danling W. Titanium carbide MXene: Synthesis, electrical and optical properties and their applications in sensors and energy storage devices: Nanomaterials Nanotechnology. 2019 Jan 17;9;1847980418824470.

[4] Bu F, Zhou W, Xu Y, Du Y, Guan C, Huang W. Recent developments of advanced micro-supercapacitors: design, fabrication and applications. Npj Flex Electron. 2020 Nov 16;4(1):1-16.

[5] Jiang Q, Lei Y, Liang H, Xi K, Xia C, Alshareef HN. Review of MXene electrochemical microsupercapacitors. Energy Storage Mater. 2020 May 1;27: 78-95.

[6] Zhang P, Wang F, Yu M, Zhuang X, Feng X. Two-dimensional materials for miniaturized energy storage devices: from individual devices to smart integrated systems. Chem Soc Rev. 2018 Oct 1;47(19):7426-51.

[7] Pech D, Brunet M, Taberna P-L, Simon P, Fabre N, Mesnilgrente F, et al. Elaboration of a microstructured inkjet-printed carbon electrochemical capacitor. J Power Sources. 2010 Feb 15;195(4):1266-9.

[8] Chmiola J, Largeot C, Taberna P-L, Simon P, Gogotsi Y. Monolithic Carbide-Derived Carbon Films for Micro-Supercapacitors. Science. 2010 Apr 23;328(5977):480-3.

[9] Pech D, Brunet M, Durou H, Huang P, Mochalin V, Gogotsi Y, et al. Ultrahigh-power micrometre-sized supercapacitors based on onion-like carbon. Nat Nanotechnol. 2010 Sep;5(9):651-4.

[10] Liang J, Mondal AK, Wang D-W, Iacopi F. Graphene-Based Planar Microsupercapacitors: Recent Advances and Future Challenges. Adv Mater Technol. 2019;4(1):1800200.

[11] Lin J, Zhang C, Yan Z, Zhu Y, Peng Z, Hauge RH, et al. 3-Dimensional Graphene Carbon Nanotube Carpet-Based Microsupercapacitors with High Electrochemical Performance. Nano Lett. 2013 Jan 9;13(1):72-8.

[12] Kurra N, Jiang Q, Nayak P, Alshareef HN. Laser-derived graphene: A three-dimensional printed graphene electrode and its emerging applications. Nano Today. 2019 Feb 1;24:81-102.

[13] Si W, Yan C, Chen Y, Oswald S, Han L, Schmidt OG. On chip, all solid-state and flexible micro-supercapacitors with high performance based on MnOx/Au multilayers. Energy Environ Sci. 2013 Oct 18;6(11):3218-23.

[14] Brezesinski T, Wang J, Tolbert SH, Dunn B. Ordered mesoporous α -MoO 3 with iso-oriented nanocrystalline walls for thin-film pseudocapacitors. Nat Mater. 2010 Feb;9(2):146-51.

[15] Du J, Cheng H-M. The Fabrication, Properties, and Uses of Graphene/ Polymer Composites. Macromol Chem Phys. 2012;213(10-11):1060-77.

[16] Kurra N, Xia C, N. Hedhili M, N. Alshareef H. Ternary chalcogenide micro-pseudocapacitors for on-chip energy storage. Chem Commun. 2015;51(52):10494-7.

[17] Naguib M, Kurtoglu M, Presser V, Lu J, Niu J, Heon M, et al.

Two-Dimensional Nanocrystals Produced by Exfoliation of Ti_3AlC_2. Adv Mater. 2011;23(37):4248-53.

[18] Naguib M, Mochalin VN, Barsoum MW, Gogotsi Y. 25th Anniversary Article: MXenes: A New Family of Two-Dimensional Materials. Adv Mater. 2014;26(7):992-1005.

[19] Anasori B, Lukatskaya MR, Gogotsi Y. 2D metal carbides and nitrides (MXenes) for energy storage. Nat Rev Mater. 2017 Jan 17;2(2):1-17.

[20] Zhan C, Sun W, Xie Y, Jiang D, Kent PRC. Computational Discovery and Design of MXenes for Energy Applications: Status, Successes, and Opportunities. ACS Appl Mater Interfaces. 2019 Jul 17;11(28):24885-905.

[21] Ghidiu M, Lukatskaya MR, Zhao M-Q, Gogotsi Y, Barsoum MW. Conductive two-dimensional titanium carbide 'clay' with high volumetric capacitance. Nature. 2014 Dec; 516(7529):78-81.

[22] Zhang Y, Wang L, Zhang N, Zhou Z. Adsorptive environmental applications of MXene nanomaterials: a review. RSC Adv. 2018 May 30;8(36):19895-905.

[23] Ibrahim Y, Mohamed A, Abdelgawad AM, Eid K, Abdullah AM, Elzatahry A. The Recent Advances in the Mechanical Properties of Self-Standing Two-Dimensional MXene-Based Nanostructures: Deep Insights into the Supercapacitor. Nanomaterials. 2020 Oct;10(10):1916.

[24] Tang H, Hu Q, Zheng M, Chi Y, Qin X, Pang H, et al. MXene–2D layered electrode materials for energy storage. Prog Nat Sci Mater Int. 2018 Apr 1; 28(2):133-47.

[25] Alhabeb M, Maleski K, Anasori B, Lelyukh P, Clark L, Sin S, et al. Guidelines for Synthesis and Processing of Two-Dimensional Titanium Carbide ($Ti_3C_2T_x$ MXene). Chem Mater. 2017 Sep 26;29(18):7633-44.

[26] Li T, Yao L, Liu Q, Gu J, Luo R, Li J, et al. Fluorine-Free Synthesis of High-Purity $Ti_3C_2T_x$ (T=OH, O) via Alkali Treatment. Angew Chem Int Ed. 2018; 57(21):6115-9.

[27] Peng C, Wei P, Chen X, Zhang Y, Zhu F, Cao Y, et al. A hydrothermal etching route to synthesis of 2D MXene (Ti_3C_2, Nb_2C): Enhanced exfoliation and improved adsorption performance. Ceram Int. 2018 Oct 15;44(15):18886-93.

[28] Li M, Lu J, Luo K, Li Y, Chang K, Chen K, et al. Element Replacement Approach by Reaction with Lewis Acidic Molten Salts to Synthesize Nanolaminated MAX Phases and MXenes. J Am Chem Soc. 2019 Mar 20;141(11):4730-7.

[29] Sun W, Shah SA, Chen Y, Tan Z, Gao H, Habib T, et al. Electrochemical etching of Ti_2AlC to Ti_2CT_x (MXene) in low-concentration hydrochloric acid solution. J Mater Chem A. 2017 Oct 24;5(41):21663-8.

[30] Yang S, Zhang P, Wang F, Ricciardulli AG, Lohe MR, Blom PWM, et al. Fluoride-Free Synthesis of Two-Dimensional Titanium Carbide (MXene) Using A Binary Aqueous System. Angew Chem Int Ed. 2018;57(47):15491-5.

[31] Khazaei M, Arai M, Sasaki T, Chung C-Y, Venkataramanan NS, Estili M, et al. Novel Electronic and Magnetic Properties of Two-Dimensional Transition Metal Carbides and Nitrides. Adv Funct Mater. 2013;23(17):2185-92.

[32] Khazaei M, Ranjbar A, Arai M, Sasaki T, Yunoki S. Electronic properties and applications of MXenes: a theoretical review. J Mater Chem C. 2017 Mar 9; 5(10):2488-503.

[33] Naguib M, Mashtalir O, Carle J, Presser V, Lu J, Hultman L, et al. Two-Dimensional Transition Metal Carbides.

ACS Nano. 2012 Feb 28;6(2): 1322-31.

[34] Zhang C (John), Anasori B, Seral-Ascaso A, Park S-H, McEvoy N, Shmeliov A, et al. Transparent, Flexible, and Conductive 2D Titanium Carbide (MXene) Films with High Volumetric Capacitance. Adv Mater. 2017;29(36): 1702678.

[35] Shahzad F, Alhabeb M, Hatter CB, Anasori B, Hong SM, Koo CM, et al. Electromagnetic interference shielding with 2D transition metal carbides (MXenes). Science. 2016 Sep 9; 353(6304):1137-40.

[36] Alnoor H, Elsukova A, Palisaitis J, Persson I, Tseng EN, Lu J, et al. Exploring MXenes and their MAX phase precursors by electron microscopy. Mater Today Adv. 2021 Mar 1;9:100123.

[37] Wang H-W, Naguib M, Page K, Wesolowski DJ, Gogotsi Y. Resolving the Structure of Ti3C2Tx MXenes through Multilevel Structural Modeling of the Atomic Pair Distribution Function. Chem Mater. 2016 Jan 12;28(1):349-59.

[38] Wang X, Shen X, Gao Y, Wang Z, Yu R, Chen L. Atomic-Scale Recognition of Surface Structure and Intercalation Mechanism of Ti3C2X. J Am Chem Soc. 2015 Feb 25;137(7):2715-21.

[39] Khazaei M, Arai M, Sasaki T, Estili M, Sakka Y. Two-dimensional molybdenum carbides: potential thermoelectric materials of the MXene family. Phys Chem Chem Phys. 2014 Apr 2;16(17) :7841-9.

[40] Shein IR, Ivanovskii AL. Graphene-like titanium carbides and nitrides Tin+1Cn, Tin+1Nn (n=1, 2, and 3) from de-intercalated MAX phases: First-principles probing of their structural, electronic properties and relative stability. Comput Mater Sci. 2012 Dec 1;65:104-14.

[41] Gao G, Ding G, Li J, Yao K, Wu M, Qian M. Monolayer MXenes: promising half-metals and spin gapless semi-conductors. Nanoscale. 1395 Feb 2;8(16):8986-94.

[42] Khazaei M, Ranjbar A, Ghorbani-Asl M, Arai M, Sasaki T, Liang Y, et al. Nearly free electron states in MXenes. Phys Rev B. 2016 May 16;93(20):205125.

[43] Lee Y, Cho SB, Chung Y-C. Tunable Indirect to Direct Band Gap Transition of Monolayer Sc2CO2 by the Strain Effect. ACS Appl Mater Interfaces. 2014 Aug 27;6(16):14724-8.

[44] Plummer G, Anasori B, Gogotsi Y, Tucker GJ. Nanoindentation of monolayer Tin+1CnTx MXenes via atomistic simulations: The role of composition and defects on strength. Comput Mater Sci. 2019 Feb 1;157:168-74.

[45] Borysiuk VN, Moachalin VN, Gogotsi Y. Molecular dynamic study of the mechanical properties of two-dimensional titanium carbides Tin+ 1Cn (MXenes). Nanotechnology. 2015 Jun;26(26):265705.

[46] Ling Z, Ren CE, Zhao M-Q, Yang J, Giammarco JM, Qiu J, et al. Flexible and conductive MXene films and nano-composites with high capacitance. Proc Natl Acad Sci. 2014 Nov 25;111(47): 16676-81.

[47] Jiang Q, Kurra N, Maleski K, Lei Y, Liang H, Zhang Y, et al. On-Chip MXene Microsupercapacitors for AC-Line Filtering Applications. Adv Energy Mater. 2019;9(26):1901061.

[48] Kim E, Lee B-J, Maleski K, Chae Y, Lee Y, Gogotsi Y, et al. Micro-supercapacitor with a 500 nm gap between MXene/CNT electrodes. Nano Energy. 2021 Mar 1;81:105616.

[49] Liu N, Gao Y. Recent Progress in Micro-Supercapacitors with In-Plane

Interdigital Electrode Architecture. Small. 2017;13(45):1701989.

[50] Xu S, Wei G, Li J, Ji Y, Klyui N, Izotov V, et al. Binder-free Ti3C2Tx MXene electrode film for supercapacitor produced by electrophoretic deposition method. Chem Eng J. 2017 Jun 1;317: 1026-36.

[51] Begines B, Alcudia A, Aguilera-Velazquez R, Martinez G, He Y, Trindade GF, et al. Design of highly stabilized nanocomposite inks based on biodegradable polymer-matrix and gold nanoparticles for Inkjet Printing. Sci Rep. 2019 Nov 6;9(1):16097.

[52] Zhang C (John), McKeon L, Kremer MP, Park S-H, Ronan O, Seral-Ascaso A, et al. Additive-free MXene inks and direct printing of micro-supercapacitors. Nat Commun. 2019 Apr 17;10(1):1795.

[53] Wu C-W, Unnikrishnan B, Chen I-WP, Harroun SG, Chang H-T, Huang C-C. Excellent oxidation resistive MXene aqueous ink for micro-supercapacitor application. Energy Storage Mater. 2020 Mar 1;25:563-71.

[54] Tang J, Yi W, Zhong X, Zhang C (John), Xiao X, Pan F, et al. Laser writing of the restacked titanium carbide MXene for high performance supercapacitors. Energy Storage Mater. 2020 Nov 1;32:418-24.

[55] Wang N, Liu J, Zhao Y, Hu M, Qin R, Shan G. Laser-Cutting Fabrication of Mxene-Based Flexible Micro-Supercapacitors with High Areal Capacitance. ChemNanoMat. 2019;5(5):658-65.

[56] Kurra N, Ahmed B, Gogotsi Y, Alshareef HN. MXene-on-Paper Coplanar Microsupercapacitors. Adv Energy Mater. 2016;6(24):1601372.

[57] Abdolhosseinzadeh S, Jiang X, Zhang H, Qiu J, Zhang C (John). Perspectives on solution processing of two-dimensional MXenes. Materials Today. 2021 Mar 15.

[58] Abdolhosseinzadeh S, Schneider R, Verma A, Heier J, Nüesch F, Zhang C (John). Turning Trash into Treasure: Additive Free MXene Sediment Inks for Screen-Printed Micro-Supercapacitors. Adv Mater. 2020;32(17):2000716.

[59] Xu S, Dall'Agnese Y, Wei G, Zhang C, Gogotsi Y, Han W. Screen-printable microscale hybrid device based on MXene and layered double hydroxide electrodes for powering force sensors. Nano Energy. 2018 Aug 1;50:479-88.

[60] Orangi J, Hamade F, Davis VA, Beidaghi M. 3D Printing of Additive-Free 2D Ti_3C_2Tx (MXene) Ink for Fabrication of Micro-Supercapacitors with Ultra-High Energy Densities. ACS Nano. 2020 Jan 28;14(1):640-50.

[61] Yang W, Yang J, Byun JJ, Moissinac FP, Xu J, Haigh SJ, et al. 3D Printing of Freestanding MXene Architectures for Current-Collector-Free Supercapacitors. Adv Mater. 2019; 31(37):1902725.

[62] Quain E, Mathis TS, Kurra N, Maleski K, Aken KLV, Alhabeb M, et al. Direct Writing of Additive-Free MXene-in-Water Ink for Electronics and Energy Storage. Adv Mater Technol. 2019;4(1): 1800256.

[63] Zhang C (John), Kremer MP, Seral-Ascaso A, Park S-H, McEvoy N, Anasori B, et al. Stamping of Flexible, Coplanar Micro-Supercapacitors Using MXene Inks. Adv Funct Mater. 2018; 28(9):1705506.

[64] Hu H, Hua T. An easily manipulated protocol for patterning of MXenes on paper for planar micro-supercapacitors. J Mater Chem A. 2017 Sep 26;5(37): 19639-48.

[65] Li P, Shi W, Liu W, Chen Y, Xu X, Ye S, et al. Fabrication of

high-performance MXene-based all-solid-state flexible microsupercapacitor based on a facile scratch method. Nanotechnology. 2018 Sep;29(44):445401.

[66] Li J, Levitt A, Kurra N, Juan K, Noriega N, Xiao X, et al. MXene-conducting polymer electrochromic microsupercapacitors. Energy Storage Mater. 2019 Jul 1;20:455-61.

[67] Chen X, Wang S, Shi J, Du X, Cheng Q, Xue R, et al. Direct Laser Etching Free-Standing MXene-MoS2 Film for Highly Flexible Micro-Supercapacitor. Adv Mater Interfaces. 2019;6(22):1901160.

[68] Peng Y-Y, Akuzum B, Kurra N, Zhao M-Q, Alhabeb M, Anasori B, et al. All-MXene (2D titanium carbide) solid-state microsupercapacitors for on-chip energy storage. Energy Environ Sci. 2016 Aug 31;9(9):2847-54.

[69] Salles P, Quain E, Kurra N, Sarycheva A, Gogotsi Y. Automated Scalpel Patterning of Solution Processed Thin Films for Fabrication of Transparent MXene Microsupercapacitors. Small. 2018;14(44):1802864.

[70] Li Q, Wang Q, Li L, Yang L, Wang Y, Wang X, et al. Femtosecond Laser-Etched MXene Microsupercapacitors with Double-Side Configuration via Arbitrary On- and Through-Substrate Connections. Adv Energy Mater. 2020;10(24):2000470.

[71] Huang X, Wu P. A Facile, High-Yield, and Freeze-and-Thaw-Assisted Approach to Fabricate MXene with Plentiful Wrinkles and Its Application in On-Chip Micro-Supercapacitors. Adv Funct Mater. 2020;30(12):1910048.

[72] Zhang L, Yang G, Chen Z, Liu D, Wang J, Qian Y, et al. MXene coupled with molybdenum dioxide nanoparticles as 2D-0D pseudocapacitive electrode for high performance flexible asymmetric micro-supercapacitors. J Materiomics. 2020 Mar 1;6(1):138-44.

[73] Huang H, Su H, Zhang H, Xu L, Chu X, Hu C, et al. Extraordinary Areal and Volumetric Performance of Flexible Solid-State Micro-Supercapacitors Based on Highly Conductive Freestanding Ti3C2Tx Films. Adv Electron Mater. 2018;4(8):1800179.

[74] Huang H, Chu X, Su H, Zhang H, Xie Y, Deng W, et al. Massively manufactured paper-based all-solid-state flexible micro-supercapacitors with sprayable MXene conductive inks. J Power Sources. 2019 Mar 1;415:1-7.

[75] Yue Y, Liu N, Ma Y, Wang S, Liu W, Luo C, et al. Highly Self-Healable 3D Microsupercapacitor with MXene–Graphene Composite Aerogel. ACS Nano. 2018 May 22;12(5):4224-32.

[76] Couly C, Alhabeb M, Aken KLV, Kurra N, Gomes L, Navarro-Suárez AM, et al. Asymmetric Flexible MXene-Reduced Graphene Oxide Micro-Supercapacitor. Adv Electron Mater. 2018;4(1):1700339.

[77] Yu C, Gong Y, Chen R, Zhang M, Zhou J, An J, et al. A Solid-State Fibriform Supercapacitor Boosted by Host–Guest Hybridization between the Carbon Nanotube Scaffold and MXene Nanosheets. Small. 2018;14(29): 1801203.

[78] Yang Q, Xu Z, Fang B, Huang T, Cai S, Chen H, et al. MXene/graphene hybrid fibers for high performance flexible supercapacitors. J Mater Chem A. 2017 Oct 31;5(42):22113-9.

[79] Li H, Li X, Liang J, Chen Y. Hydrous RuO2-Decorated MXene Coordinating with Silver Nanowire Inks Enabling Fully Printed Micro-Supercapacitors with Extraordinary Volumetric Performance. Adv Energy Mater. 2019;9(15):1803987.

[80] Wang Y, Wang X, Li X, Bai Y, Xiao H, Liu Y, et al. Scalable fabrication of polyaniline nanodots decorated MXene film electrodes enabled by viscous functional inks for high-energy-density asymmetric supercapacitors. Chem Eng J. 2021 Feb 1;405:126664.

[81] Jiao S, Zhou A, Wu M, Hu H. Kirigami Patterning of MXene/Bacterial Cellulose Composite Paper for All-Solid-State Stretchable Micro-Supercapacitor Arrays. Adv Sci. 2019;6(12):1900529.

[82] Shao W, Tebyetekerwa M, Marriam I, Li W, Wu Y, Peng S, et al. Polyester@MXene nanofibers-based yarn electrodes. J Power Sources. 2018 Aug 31;396:683-90.

[83] Zhang J, Seyedin S, Gu Z, Yang W, Wang X, Razal JM. MXene: a potential candidate for yarn supercapacitors. Nanoscale. 2017 Dec 7;9(47):18604-8.

[84] He N, Liao J, Zhao F, Gao W. Dual-Core Supercapacitor Yarns: An Enhanced Performance Consistency and Linear Power Density. ACS Appl Mater Interfaces. 2020 Apr 1;12(13):15211-9.

Section 5

Hybrid Capacitors

Chapter 7

Supercapacitor Supported by Nickel, Cobalt and Conducting Polymer Based Materials: Design Techniques and Current Advancement

Satish P. Mardikar, Sagar D. Balgude and Santosh J. Uke

Abstract

The recent advanced electronic appliances demand special high power devices with lightweight, flexible, inexpensive, and environment friendly in nature. In addition, for many industrial and automotive applications, we need energy storage systems that can store energy in a short time and deliver an intense pulse of energy for long duration. Till date the Li-ion battery is the only choice for fulfilling all our energy storage demands. However, the high cost, limited availability and non-environmental nature of electrodes and electrolyte material of Li-ion battery limits its applicability. Hence, the world demands an alternative replacement for the Li-ion battery. In this regard, the supercapacitor is one of the most emerging and potential energy storage devices. The electrode plays an important role in supercapacitors. The nickel and cobalt based oxide, hydroxides, and their composites with conducting polymer are promising and highly appreciated electrode materials for supercapacitors. This chapter covers the recent advances in supercapacitors supported by nickel, cobalt and conducting polymer based materials and their applications predominantly described in the recent literature. Recent advances are reviewed including new methods of synthesis, nanostructuring, and self-assembly using surfactant and modifiers. This chapter also covered the applications of supercapacitors in powering the light weight, flexible and wearable electronics.

Keywords: supercapacitors, mixed ternary metal oxides, nanostructured, $NiCo_2O_4$, energy density, etc.

1. Introduction

Supercapacitor (SCp) is also known as ultracapacitor. SCP is the advanced electrochemical energy storage device. At present, the lithium ion battery (LiBs), lead acid battery and SCP are the major available energy storage systems. Over the other energy storage systems, the SCP is stands out to be a promising energy storage device with very attractive properties such as high specific capacitance, high power density, moderate energy density, good cyclic stability, low cost, environmental friendly nature, etc. SCP has been utilized in various electrical applications viz.

hybrid vehicles, power backup, military services, and portable electronic devices like laptops, mobile phones, roll-up displays, electronic papers, etc. [1–3].

The performance of SCP is strongly depends on types electrode materials i.e. active material used in supercapacitor. Based on the type of active material used and process energy storage, the SCP can be divided into three main categories, including pseudocapacitors (PCs), electric double-layer capacitors (EDLCs) and hybrid capacitors [4–6]. In PCs metal oxides, metal hydroxides and conducting polymers are employed as active material. On the other hand, carbon base materials such carbon nanotubes, graphene and carbon black, etc. are used as active material in EDLCs employed. Likewise, the used of combination of metal oxide, conducting polymer and carbon based material as active material results hybrid capacitor. The PCs delivered high specific capacitance, high energy density than EDLCs, but demonstrates poor power density and cycle stability. Nevertheless, owing to the high active surface area, the EDLCs delivered high specific capacitance but suffer poor energy density than PCs [4]. The hybrid capacitors retain the advantage of both PCs and EDLCs and hence they delivered high specific capacitance, high energy density, and large cycle life [7, 8]. Moreover, the performance of supercapacitors is equally rely on different aspect of active material viz. quality, electric conductivity, material, size, porosity, synthesis method, etc. More specifically, the synthesis method can bring many attractive advantages in active material for extraordinary electrochemical performance of supercapacitor. Therefore, synthesis of active materials with high porosity, stable performance and good electrical conductivity has a very wide research potential.

Recently, to enhance the energy density, cycle life and electrochemical performance of the supercapacitors, the use of electrode material with desired structure with uniform porosity is one of the appealing strategies. From many decades, the nanostructured single transition metal oxides such as RuO_2 [9], MnO_2 [10], CeO_2 [11], Fe_2O_3 [12], Fe_3O_4 [13], Co_3O_4,, Mn_3O_4,, etc., and the nanostructured mixed ternary metal oxide (TMOs) such as $ZnFe_2O_4$, $NiFe_2O_4$ $CuFe_2O_4$, $CoFe_2O_4$, $MnCo_2O_4$ $ZnCo_2O_4$, $NiCo_2O_4$, and etc., and conducting polymers such as, polyaniline (PANi), polypyrrole (Ppy), polythiophene (Pth), etc. has been extensively used as active material for all types of supercapacitors. Out of the different materials used as the electrodes for SCp applications. The TMOs are highly studied and excessively used as active material in all types of supercapacitors. The mixed TMOs are also called the spinel metal oxide. The spinel TMOs have the general formula AB_2O_4. In AB_2O_4, the cubic crystal structure of TMOs consists of closely packed O^{2-} anions and an octahedral and tetrahedral space of the lattice occupied by the transition metal cations A and B, respectively. Due to this closed packed structure, the mixed TMOs show the extraordinary characteristics over single metal oxides, such as two order higher electrical conductivity, superior electrochemical performance and excellent stability over single metal oxides. Moreover, the recent research reports shows the TMOs have better structural advantages and higher surface area and porosity [14]. More specifically, the TMOs show low cost, natural abundance, low toxicity and environmental friendly nature. Hence, TMOs have drowned more research attention in recent years. In addition, the extraordinary electrochemical performance of TMOs in solid as well liquid electrolyte makes it a promising and potential candidate as electrode material for PCs. The various mixed TMOs viz. $ZnFe_2O_4$, $NiFe_2O_4$ $CuFe_2O_4$, $CoFe_2O_4$, $MnCo_2O_4$ $ZnCo_2O_4$, $NiCo_2O_4$, etc. has been utilized as electrode material for PCs. Out of the different TMOs the nickel and cobalt based TMOs have gained more research attention as electrode material in supercapacitors due to their attracting properties such as low cost, natural abundance, low toxicity and environmental friendly nature. More specifically, these materials show variable structures, diverse morphologies, high specific surface and uniform porosity and outstanding electrical conductivity. The $NiCo_2O_4$ demonstrated high electrical conductivity due to the presence of Ni in it. Whereas, Co enhances the electrochemical activity of oxides, further, the synergistic effect among Ni in Co offers high electrical conductivity with an excellent electrochemical behavior in supercapacitors [12]. The $NiCo_2O_4$ demonstrated a high theoretical capacity [15]. The nickel and cobalt based TMOs show diverse morphologies, this includes various nanostructures ranging from 0 to 3 D architectures viz. quantum dots, nanowires, nanosheets, platelets like nanoparticles, porous network like framework, coral- like porous crystals, ordered mesoporous particles, urchin-like microstructures and urchin-like nanostructures. Till date many recent attractive reviews have presented recent development in mixed TMOs as electrode material for SCs [7, 16–19]. We recommend few of them for readers who are new to this field of energy research.

In the present chapter we provide the recent advancements in synthesis of nanostructured nickel and cobalt base mixed TMOs and their composites with conducting polymer based materials as electrode material for supercapacitors predominantly described in the recent literature. Moreover, here our special emphasis will be on new methods of synthesis, nanostructuring, and self-assembly using surfactant and modifiers. In addition, we provide a summary of structural and morphological advancements regarding the electrochemical properties of supercapacitors. Finally, we link our discussion to the recent applications in powering the light weight, flexible and wearable electronics real world applications.

2. Synthesis of nickel and cobalt base mixed TMOs

Compared to the micro sized the nanostructured cobalt and nickel based TMOs show higher specific capacitance and long cycle life. Therefore many recent research strategies have drowned to synthesized nanosize mixed cobalt and nickel based TMOs. The various synthesis methods for synthesis of nanostructured cobalt and nickel based mixed TMOs viz., hydrothermal method sol–gel, thermal evaporation method, chemical bath deposition, electrodeposition, oil/water interfacial self-assembly strategy, etc. have been extensively reported in literature. Hydrothermal method is one of the excessively adopted synthesis methods for hierarchical nanostructure synthesis. This method is cost effective, simple, and easy to scale-up at room temperature. This method is mostly used for fine tuning the morphology and controlling the size of nanostructures. Agglomeration of $NiCo_2O_4$ results in low electrical conductivity and decreases the specific capacitance and cycle life of SCp [20]. Therefore, to enhance the electrical conductivity the use of high surface area with high porosity conductive substrates are highly recommended. These substrates enhance the contact between electrode and electrolyte and allowed more electrolyte ions penetration in active material. The various conductive substrates such as textiles, sponges, carbon clothes, carbon fibers, conventional paper, cables, etc. are used as substrates to fabricate SCs. Such conductive substrates are advantageous for enhancing the electrochemical performance via providing short diffusion path, high electrical conductivity, ample electroactive sites [21]. In this regard, Yang *et al.* synthesized the nanoneedle arrays of on filter carbon paper substrate via facial hydrothermal synthesis method. For fabrication the filter carbon paper submerged into the precursor of $NiCo_2O_4$ followed by calcination in the Argon atmosphere in the range of 250-400°C for 2 hours. Using this approach they have reported urchin like and nanoneedle arrays and further adopted this nanostructure for SCp applications [20]. The synthesis parameters like reaction temperature reaction temperature and reaction time controls the structures. Further, the calcination temperature after synthesis plays crucial role in improving the surface morphology, specific surface area, porosity, etc. [22]. Siwatch *et al.* [22] have reported the formation of $NiCo_2O_4$ quantum dots via hydrothermal synthesis and studied the effect of synthesis parameters like reaction temperature and time, and calcination temperature on the morphology of $NiCo_2O_4$ quantum dots. Further, the highly porous flower-like structure of $NiCo_2O_4$ quantum dots obtained at the calcination temperature 300° C is highly useful for SCp applications. Lu *et al.* reported the synthesis of mesoporous $NiCo_2O_4$ via reagents assisted hydrothermal method and studied the effect of reagent cetyltrimethylammonium bromide (CTAB) on morphology and electrochemical behavior of $NiCo_2O_4$ for SCp applications. Moreover, the reagent during synthesis enhanced the specific surface area and charge transport of $NiCo_2O_4$. As a result, the cyclic stability, rate performance and specific capacitance of $NiCo_2O_4$ quantum dots based in asymmetric SCp found to be increased [23]. Binder used during the fabrication of electrode increase the electrode resistance which further decreases the electrochemical performance and cycle life of SCp. Therefore recently binder free fabrication approaches such as direct growth on conductive substrate is more popular. Furthermore, over the conventional substrates the direct growth on three dimensional (3D) conductive substrate offers many advantages such as shorten the diffusion path, healthy synergy between the active material and electrolyte, provides ample electroactive site, etc. which further help to enhance the electrochemical performance of SCp. For example, Yang *et al.* [24] directly grown $NiCo_2O_4$ on gelatin-based carbonenickel foam (3D) by facial hydrothermal method followed by calcination at 350°C for 2 hrs under an Argon atmosphere.

The morphology of $NiCo_2O_4$ is reported to be nanoflower-like. The fabricated 3D electrode provides fast ions and electrons transfer rate and enhances the electroactive surface area of the $NiCo_2O_4$ via forming a complex 3D network. In addition, the nickel foam as substrate adds the electric conductivity whereas the gelatin based carbon on the nickel foam provides high surface area for uniform growth of NiCo2O4 during synthesis. In our previous study, we have reported the synthesis of nanostructured $NiCo_2O_4$ via surfactant assisted hydrothermal method and studied the effect surfactant and reaction parameters on the morphology of nanostructured $NiCo_2O_4$. From this synthesis, we got two distinct morphologies viz. platelet-like and nanorod-like using surfactants TEA ethoxylate and polyethylene glycol (PEG), respectively. We further used this nanostructured $NiCo_2O_4$ for SCp applications [18].

In addition to the hydrothermal method, the combustion method is one of the simple and easy to scalable synthesis methods. Over the hydrothermal method the combustion method does not requires Teflon-lined stainless steel autoclaves and centrifuge for product washing, is less time consuming and provides high phase purities. This regard, Kumar *et al.* reported the growth of $NiCo_2O_4$ on conductive substrate nickel foam using combustion method. For the synthesis, honeycomb-like $NiCo_2O_4$ the nitrate and glycine used as oxidizer and fuel, respectively [6].

For enhancing the surface area and porosity and electrochemical activities of $NiCo_2O_4$, the formation of composites of $NiCo_2O_4$ with carbon based material is one of the appealing strategies, for example, $NiCo_2O_4$/CNT, $NiCo_2O_4$/MWCNT, $NiCo_2O_4$/graphene, $NiCo_2O_4$/reduced graphene oxides (r-GO), etc. demonstrated to be a potential candidates for SCp applications. Carbon base material viz. CNT, MWCNT, graphene, r-GO, etc. provide excellent flexibility, high specific surface areas, remarkable electrical conductivity, good thermal and chemical stability [5, 25–27]. For example, Li *et al.* [25]. reported the synthesis of $NiCo_2O_4$/CNT composites, and studied the structure formation of $NiCo_2O_4$ with and without surfactant for supercapacitive applications. The nanoflakes and nanocorn like morphology for $NiCo_2O_4$ is obtained by using surfactant sodium dodecyl sulfate. Pathak *et al.* synthesized $NiCo_2O_4$ and $NiCo_2O_4$@ MWCNT composite using facile hydrothermal method.

$NiCo_2O_4$@ MWCNT demonstrated superior electrochemical performance and demonstrated a good electrode for SCp applications. In addition, using density functional they reveal the enhanced density of states near the Fermi level and increased quantum capacitance of the $NiCo_2O_4$ @SWCNT is one of the important reasons for high specific capacitance, high power density and energy density [28]. The PCs use reversible fast faradaic reactions to store electrical charges, which allow them to achieve higher capacitance by at least one order of magnitude than those obtained by EDLCs. Materials sustaining such redox reactions on their surfaces include, for example, conducting polymers and transition metal oxides.

3. Applications of Ni and Co based metal oxides and their composites

3.1 Pseudocapacitor (PCs)

Recently, PCs received considerable attention due to the one order higher capacitance, higher volumetric capacitance, higher energy density and use of low cost and easily synthesized active material than EDLCs. [28–30]. For example, Eskandari *et al.*. fabricated $NiCo_2O_4$ and its composite with PANi and MWCNTs and reduced graphene oxide r-GO and studied their SCp performance in 3 M KOH. Out of the different composites the $NiCo_2O_4$/PANi demonstrated superior performance and exhibited specific capacitance of 1760 Fg^{-1} (900 F/g and 734 F/g for $NiCo_2O_4$/ MWCNTs and NiCo2O4/r-GO, respectively) at current density of 1 Ag^{-1}, respectively. The highest specific capacitance in $NiCo_2O_4$/PANi is due to supplementary conductive pathways provided by PANi and synergistic effect of the rooted pseudo-reaction. Moreover the composite $NiCo_2O_4$/MWCNTs shows stable cycle is life and demonstrated to be best retention over all other composites as 89% over 2000 charge discharge cycles. Composite $NiCo_2O_4$ /r-GO also exhibited good cycle performance and shows retention in specific capacitance of 87% over 2000 charge discharge cycles. In addition, the pristine $NiCo_2O_4$ shows higher retention than $NiCo_2O_4$/PANi, i.e. 70% and 73%, respectively. The highest cyclic stability in $NiCo_2O_4$/MWCNTs is due to the high electrical conductivity and high mechanical strength of MWCNTs. In fact, the good cyclic performance in $NiCo_2O_4$/r-GO is the results of higher electrical conductivity, high surface morphology and good mechanical strength than PANi and pristine $NiCo_2O_4$ [31] Moreover, the representative Ni and Co based material and their performance in PCs are summarized in **Table 1**.

3.2 Hybrid capacitors

Even if the Ni and Co based TMOs are advantageous for SCp applications, however, in the long cycling process the rapid degradation of $NiCo_2O_4$ electrode materials is the major obstacle among the commercialization of $NiCo_2O_4$ based SCp. By increasing the electrical conductivity of $NiCo_2O_4$ this hurdle can be minimized and the higher rate capabilities can be attained. Therefore, from the last two decades, researchers devoted more efforts to enhance the electrical conductivity of $NiCo_2O_4$, this includes fabrication of hybrid composite with other conducting electrode materials, viz. carbon based material (CNts, SWCNts, MWCNts, activated carbon, doped and undoped reduced graphene oxides, etc.), conducting polymers, etc. In addition, recent formation of composite of $NiCo_2O_4$ with other mixed TMOs has gain enormous attention. For example, Mary *et al.* reported the fabrication of $NiCo_2O_4$ and $ZnCo_2O_4$ composites and studied their morphology dependent electrochemical behavior for hybrid SCp applications. In addition, the hybrid SCp

Sr. No.	Material	Method of synthesis	Electrolyte	Voltage window (V)	Specific capacitance (Fg^{-1}) at current density-scan rate	Energy density $Whkg^{-1}$	Retention of capacitance at (current density) (cycle numbers)	Reference
1.	$NiCo_2O_4$	Hydrothermal	1 M Na_2SO_4	0–0.6 V	479/ 5 mVs^{-1}	21.3	87.21% (5000)	[30]
2.	$NiCo_2O_4$	Hydrothermal	1 M Na_2SO_4	0–0.4 V	320 0.1 mVs^{-1}	16.1	95.34%, (1000)	[32]
3.	$NiCo_2O_4$ /NiCo2S4	Molecular design	—	0.1–0.6 V	1296 1 Ag^{-1}	44.8	93.2% (6000) 10 Ag^{-1},	[1]
4.	$NiCo_2O_4$	Hydrothermal	3 M KOH	0.0–0.6 V	3143 2 mVs^{-1}	56	48% (5000) 10 Ag^{-1},	[33]
5.	$NiCo_2O_4$ @α-Co(OH)2 nanowires	Hydrothermal	2 M KOH	−0.2 -0.5 V	1298-1 Ag^{-1}	39.7	83% (5000) 2 Ag^{-1}	[34]
6.	Mesoporous $NiCo_2O_4$ nano-needles	Hydrothermal	—	0.0–0.5 V	1410 Fg^{-1}-1 Ag^{-1}		94.7% (3000) 20 Ag^{-1}	[35]
7.	$NiCo_2O_4$ nanosheets	Solvothermal	—	—	2690 Fg^{-1}	52.6	80.9% (3,000) 20 mA cm^{-2}.	[36]
8.	$NiCo_2O_4$ PANi	Hydrothermal and in-situ polymerization	6 M KOH	0–0.5 V	3108 1 mA·cm^{-2}.	77.57	96.1% (1000)	[37]
9.	$NiCo_2O_4$ nanoneedles	Hydrothermal via annealing approach	1 M KOH	0–0.7 V	1076 0.5 Ag^{-1}	30.5	14% (1000) 10 Ag^{-1}	[38]
10.	$NiCo_2O_4$ nanoneedles	Pulsed laser ablation	3 M KOH	0–0.6 V	1650 Fg^{-1} 1 Ag^{-1}	56.7	91.78%, (12,000) 10 Ag^{-1}	[39]

Table 1.
Overview of representative Ni and Co based material and their performance in PCs.

Sr.No.	Material	Method of synthesis	Electrolyte	Voltage window (V)	Specific capacitance (Fg^{-1}) at current density/ scan rate	Energy density $Whkg^{-1}$	Retention of capacitance (cycle numbers) at current density	Reference
1	$NiCo_2O_4$ CNT	Hydrothermal	2 M KOH	−0.1- 0.5	574.3 0.5 Ag^{-1}	—	111.5% (1000)	[25]
2	$NiCo_2O_4$ @MWCNT	Hydrothermal	0.5 M K_2SO_4	—	374 2 Ag^{-1}	95	74.85% (3000)	[28]
3	3D $NiCo_2O_4$ /MWCNT	Sol–gel	2 M KOH	0–0.5 V	1010 0.1 Ag^{-1}	37.7	83.4% (2000) 2 Ag^{-1}	[40]
4	Ordered Mesoporous Carbon/$NiCo_2O_4$	co-precipitation	6 mol· L^{-1} KOH	0–0.6 V	577.0 1 Ag^{-1}	—	92.7%. (2000) 2 Ag^{-1},	[41]
5	$NiCo_2O_4$- nanoporous carbon.	Chemical	1 M KOH	−0.2-0.6 V	89 0.1 - Ag^{-1}	28	85% (2000)	[42]
6	Mesoporous carbon - $NiCo_2O_4$	hydrothermal followed by calcination	3 M KOH	−0.45-0.45	204.28 1 - Ag^{-1}	5.75	90.35% (3000) 20 Ag^{-1}	[43]
7	Hallow bamboo-shaped $NiCo_2O_4$	Template	6 M KOH	0.0–0.6 V	680.1C g^{-1} 1 Ag^{-1}	59.82	99.7% (5000) 10 Ag^{-1}.	[44]
8.	rGO- $NiCo_2O_4$ quantum dots	Chemical	1 M Na2SO4	0.0–1.6 V	265 0.73 Ag^{-1}	47	69% (1000)	[45]
9.	Oxygen-vacancy-rich $NiCo_2O_4$/nitrogen-deficient graphitic carbon nitride hybrids	Chemical	6 M KOH	0.0–0.6 V	1998 2 Ag^{-1}	70.22	95.22% (5000)	[46]
10	$NiCo_2O_4$@Ppy/CC	Hydrothermal	2 M KOH	0.0–0.5 V	155.4 mAh g^{-1} 1 mA cm^{-1}	22.3	71% (8000) 10 mA cm^{-2}	[47]

Table 2.
Overview of representative Ni and Co based material and their performance hybrid supercapacitor.

Sr. No.	Material	Method of synthesis	Electrolyte	Voltage window (V)	Specific capacitance (Fg^{-1}) at current density/ scan rate	Energy density $Whkg^{-1}$	Retention of capacitance (cycle numbers) at current density	Reference
1	$NiCo_2O_4$/C composites // activated carbon (AC)	Hydrothermal	6 M KOH	0–1.4 V	995.2 10 Ag^{-1},	20.87	83.04% (5000) 100 mVs^{-1}	[20]
2	$NiCo_2O_4$ quantum dots // reduced graphene oxide (rGO)	Hydrothermal	1 M Na_2SO_4	0–2.4 V	362, 0.5 Ag^{-1}.	69.5	86% (1000)	[22]
3	$NiCo_2O_4$ //AC	Hydrothermal	—	0–0.5 V	153.2 (1 Ag^{-1}),	22.5	97.1% (1000) 1 Ag^{-1}	[23]
4	$NiCo_2O_4$/ gelatin-based carbonenickel foam // AC	Hydrothermal		0–0.5 V	1416 (1 Ag^{-1})	48.6	88.5% (10,000) 5 Ag^{-1}.	[24]
5	$NiCo_2O_4$@Ni foam// sugar-derived carbon (SC)	Combustion	6 M KOH	0–0.5 V	169, 1.5 Ag^{-1}	48	96.5% (3000)	[6]
6.	MWCNTs intermingled $NiCo_2O_4$// Cu_2WS_4	Hydrothermal	3 M KOH	0–0.5 V	116 mAh g^{-1}, 1 Ag^{-1}	87	~88%, (10,000)	[5]
7.	$NiCo_2O_4$ nanoparticles and nanowires	Hydrothermal and wet chemical	1 M Na_2SO_4	0–1.6 V	1066.03 Ag^{-1}	59.56	77% (5000) 6.66 Ag^{-1}	[48]
8	$NiCo_2O_4$- carbon nanofiber	—	2 M KOH	—	991.96 5 Ag^{-1}	37.23	97.02% (3000) 30 Ag^{-1}	[15]
9	$NiCo_2O_4$/ r-GO	Hydrothermal	1 M KOH	0.0–1.6 V	702 0.5 Ag^{-1}	—	92% (1000) 4 mA cm^{-1}	[49]
10	$NiCo_2O_4$/ rGO	Co-precipitation	1 M KOH	0.0–0.6 V	1380 1 Ag^{-1}	—	90% (1000) 5 Ag^{-1} after	[50]

Table 3.
Overview of representative Ni and Co based material and their performance asymmetric supercapacitor.

fabricated using the $NiCo_2O_4$ and $ZnCo_2O_4$ composite and nitrogen doped activated carbon. The high surface area and uniform porosity of activated carbon in hybrid SCp enhances the capacitance via enabling the more electrolyte ions into active material. Interestingly, $NiCo_2O_4$ @ $ZnCo_2O_4$ composite shows high specific capacitance of 236 C g^{-1} at a current density of 1 A g^{-1}. Moreover, the aforementioned hybrid SCp results in high energy density of 101.6 $Whkg^{-1}$ and high retention in capacitance at 78.5% over 12000 charge–discharge cycles. Moreover, the representative Ni and Co based material and their performance in hybrid SCp are summarized in **Table 2**.

3.3 Asymmetric capacitors

The symmetrical SCp limits their specific capacitance due to narrow potential windows. Moreover, the use of aqueous base liquid electrolyte in symmetrical SCp decreases the specific capacitance energy density and cycle life. To overcome such drawback the fabrication of SCp with two different kinds of active material based electrode is demonstrated to be an effective strategy. The SCp fabricated using two different electrodes is termed an asymmetric supercapacitor. In asymmetric SCp positive electrode is fabricated using metal oxide base material, while negative electrode is fabricated by carbon based material. The combination of different active materials in a single device with higher operating potential result in higher the specific capacitance and energy density [40]. However, aqueous-based symmetric supercapacitors suffer from narrow potential windows, due to the limitation of the water decomposition. Therefore, an effective way is to construct asymmetric supercapacitor, which consists of two kinds of electrode materials, for instance positive electrode having pseudocapacitive nature and negative electrode having electric double layer capacitance with higher operating potential, for obtaining higher energy density [19, 20]. In the case of positive electrode materials, transition metal oxide based nanoparticles, conducting polymers based materials have been widely utilized, which exhibits pseudocapacitance as well as reversible redox Faradaic reaction. As negative electrode materials, carbon based materials like carbon nanotubes (CNT), graphene oxide (GO), activated carbon, and mesoporous carbon materials displaying electric double layer capacitance have been used. Among the carbon allotropes, mesoporous carbons have been extensively used as negative electrode material due to its high surface area and good electrical conductivity. For more understanding the recent advancements in $NiCo_2O_4$ and their composites and their performance in asymmetric SCp are summarized in **Table 3**.

4. Conclusions and outlooks

With ever increasing energy demands, day by day the SCp gaining much interest as an energy storage device. From a many years, the nickel and cobalt based TMOs and their composites have been studied and successively employed as an active material in all types of SCp. The nanostructured $NiCo_2O_4$ is low cost, in abundance, environmentally friendly in nature and has high electrical conductivity. In addition, due to the enhanced mobility of charge carriers the nanostructured $NiCo_2O_4$ demonstrated to be higher electrochemical performance than the single metal oxides. In this regard, the recent advances in synthesis of pristine $NiCo_2O_4$ and their composites with diverse morphologies and their applications for electrochemical performance in all types of SCp have been summarized in this chapter. Out of the different synthesized methods used for synthesis nanostructured $NiCo_2O_4$, the hydrothermal method is found to be excessively used. Moreover, the hydrothermal method is demonstrated to be more advantageous for the synthesis of diverse morphologies ranging from 0 D to 3 D, and resulted in high specific surface area and uniform porosity. The pristine nanostructured $NiCo_2O_4$ has many limitations for its commercial supercapacitor applications. Therefore, advanced strategies like synthesis of hierarchical nanostructures of $NiCo_2O_4$ and the fabrication of composite with other mixed TMOs, carbon based material and conducting polymers can enhance the specific capacitance, energy density and rate capability of Ni and Co based supercapacitors.

Conflict of interest

The authors declare no conflict of interest.

Author details

Satish P. Mardikar[1], Sagar D. Balgude[2] and Santosh J. Uke[3]*

1 Smt. R.S. College, SGB Amravati University, Amravati, India

2 Department of Chemistry, D.Y. Patil University, Pune, India

3 JDPS College, SGB Amravati University, Amravati, India

*Address all correspondence to: phyuke@gmail.com

References

[1] Y. Liu, Z. Wang, Y. Zhong, M. Tade, W. Zhou, Z. Shao, Molecular Design of Mesoporous NiCo2O4 and NiCo2S4 with Sub-Micrometer-Polyhedron Architectures for Efficient Pseudocapacitive Energy Storage, Advanced Functional Materials. 27 (2017) 1701229. doi:10.1002/adfm.201701229.

[2] S.J. Uke, V.P. Akhare, D.R. Bambole, A.B. Bodade, G.N. Chaudhari, Recent advancements in the cobalt oxides, manganese oxides, and their composite as an electrode material for supercapacitor: A review, Frontiers in Materials. 4 (2017) 21.

[3] S.J. Uke, G.N. Chaudhari, Y. Kumar, S.P. Mardikar, Tri-Ethanolamine-Ethoxylate assisted hydrothermal synthesis of nanostructured MnCo2O4 with superior electrochemical performance for high energy density supercapacitor application, Materials Today: Proceedings. (2020).

[4] J.-S.M. Lee, M.E. Briggs, C.-C. Hu, A.I. Cooper, Controlling electric double-layer capacitance and pseudocapacitance in heteroatom-doped carbons derived from hypercrosslinked microporous polymers, Nano Energy. 46 (2018) 277-289.

[5] V. Sharma, S.J. Kim, N.H. Kim, J.H. Lee, All-solid-state asymmetric supercapacitor with MWCNT-based hollow NiCo2O4 positive electrode and porous Cu2WS4 negative electrode, Chemical Engineering Journal. 415 (2021) 128188.

[6] D.R. Kumar, K.R. Prakasha, A.S. Prakash, J.-J. Shim, Direct growth of honeycomb-like NiCo2O4@Ni foam electrode for pouch-type high-performance asymmetric supercapacitor, Journal of Alloys and Compounds. 836 (2020) 155370. doi:10.1016/j.jallcom.2020.155370.

[7] L. Liu, H. Zhao, Y. Lei, Review on nanoarchitectured current collectors for pseudocapacitors, Small Methods. 3 (2019) 1800341.

[8] A. Ehsani, A.A. Heidari, H.M. Shiri, Electrochemical pseudocapacitors based on ternary nanocomposite of conductive polymer/graphene/metal oxide: An introduction and review to it in recent studies, The Chemical Record. 19 (2019) 908-926.

[9] M.-Y. Chung, C.-T. Lo, High-performance binder-free RuO2/electrospun carbon fiber for supercapacitor electrodes, Electro chimica Acta. 364 (2020) 137324.

[10] Y. Kumar, S. Chopra, A. Gupta, Y. Kumar, S.J. Uke, S.P. Mardikar, Low temperature synthesis of MnO2 nanostructures for supercapacitor application, Materials Science for Energy Technologies. 3 (2020) 566-574.

[11] H. Xing, G. Long, J. Zheng, H. Zhao, Y. Zong, X. Li, Y. Wang, X. Zhu, M. Zhang, X. Zheng, Interface engineering boosts electrochemical performance by fabricating CeO2@ CoP Schottky conjunction for hybrid supercapacitors, Electrochimica Acta. 337 (2020) 135817.

[12] S. Cheng, Y. Zhang, Y. Liu, Z. Sun, P. Cui, J. Zhang, X. Hua, Q. Su, J. Fu, E. Xie, Energizing Fe2O3-based supercapacitors with tunable surface pseudocapacitance via physical spatial-confining strategy, Chemical Engineering Journal. 406 (2021) 126875.

[13] C. Liu, Q. Li, Q. Zhang, B. He, P. Man, Z. Zhou, C. Li, L. Xie, Y. Yao, Surface-functionalized Fe2O3 nanowire arrays with enhanced pseudocapacitive performance as novel anode materials for high-energy-density fiber-shaped

asymmetric supercapacitors, Electrochimica Acta. 330 (2020) 135247.

[14] D. Chen, D. Pang, S. Zhang, H. Song, W. Zhu, J. Zhu, Synergistic coupling of NiCo2O4 nanorods onto porous Co3O4 nanosheet surface for tri-functional glucose, hydrogen-peroxide sensors and supercapacitor, Electrochimica Acta. 330 (2020) 135326. doi:10.1016/j.electacta.2019.135326.

[15] G.M. Tomboc, H. Kim, Derivation of both EDLC and pseudocapacitance characteristics based on synergistic mixture of NiCo2O4 and hollow carbon nanofiber: An efficient electrode towards high energy density supercapacitor, Electrochimica Acta. 318 (2019) 392-404. doi:10.1016/j.electacta.2019.06.112.

[16] H. Wang, Y. Zhong, J. Ning, Y. Hu, Recent advances in the synthesis of non-carbon two-dimensional electrode materials for the aqueous electrolyte-based supercapacitors, Chinese Chemical Letters. (2021). doi:10.1016/j.cclet.2021.04.025.

[17] M.A.A. Mohd Abdah, N.H.N. Azman, S. Kulandaivalu, Y. Sulaiman, Review of the use of transition-metal-oxide and conducting polymer-based fibres for high-performance supercapacitors, Materials & Design. 186 (2020) 108199. https://doi.org/10.1016/j.matdes.2019.108199.

[18] M. Bigdeloo, E. Kowsari, A. Ehsani, A. Chinnappan, S. Ramakrishna, R. AliAkbari, Review on innovative sustainable nanomaterials to enhance the performance of supercapacitors, Journal of Energy Storage. 37 (2021) 102474. doi:10.1016/j.est.2021.102474.

[19] S.A. Delbari, L.S. Ghadimi, R. Hadi, S. Farhoudian, M. Nedaei, A. Babapoor, A. Sabahi Namini, Q.V. Le, M. Shokouhimehr, M. Shahedi Asl, M. Mohammadi, Transition metal oxide-based electrode materials for flexible supercapacitors: A review, Journal of Alloys and Compounds. 857 (2021) 158281. doi:10.1016/j.jallcom.2020.158281.

[20] G. Yang, S.-J. Park, Facile hydrothermal synthesis of NiCo2O4-decorated filter carbon as electrodes for high performance asymmetric supercapacitors, Electrochimica Acta. 285 (2018) 405-414. doi:10.1016/j.electacta.2018.08.013.

[21] D.P. Dubal, J.G. Kim, Y. Kim, R. Holze, C.D. Lokhande, W.B. Kim, Supercapacitors based on flexible substrates: an overview, Energy Technology. 2 (2014) 325-341.

[22] P. Siwatch, K. Sharma, S.K. Tripathi, Facile synthesis of NiCo2O4 quantum dots for asymmetric supercapacitor, Electrochimica Acta. 329 (2020) 135084. doi:10.1016/j.electacta.2019.135084.

[23] Z. Lu, D. Xuan, D. Wang, J. Liu, Z. Wang, Q. Liu, D. Wang, Y.-Y. Ye, Z. Zheng, S. Li, Reagents-assisted hydrothermal synthesis of NiCo2O4 nanomaterial as electrode for high-performance asymmetric supercapacitor, New J. Chem. (2021). doi:10.1039/D1NJ00268F.

[24] G. Yang, S.-J. Park, Nanoflower-like NiCo2O4 grown on biomass carbon coated nickel foam for asymmetric supercapacitor, Journal of Alloys and Compounds. 835 (2020) 155270. doi:10.1016/j.jallcom.2020.155270.

[25] Y. Li, L. Zheng, W. Wang, Y. Wen, Controllable Synthesis of NiCo2O4/CNT Composites for Supercapacitor Electrode Materials, Int. J. Electrochem. Sci. 15 (2020) 11567-11583.

[26] B. Asbani, K. Robert, P. Roussel, T. Brousse, C. Lethien, Asymmetric micro-supercapacitors based on electrodeposited Ruo2 and sputtered VN films, Energy Storage Materials. 37 (2021) 207-214.

[27] N. Yadav, S.A. Hashmi, Hierarchical porous carbon derived from eucalyptus-bark as a sustainable electrode for high-performance solid-state supercapacitors, Sustainable Energy & Fuels. 4 (2020) 1730-1746.

[28] M. Pathak, J.R. Jose, B. Chakraborty, C.S. Rout, High performance supercapacitor electrodes based on spinel NiCo2O4@ MWCNT composite with insights from density functional theory simulations, The Journal of Chemical Physics. 152 (2020) 064706.

[29] M. Eskandari, R. Malekfar, D. Buceta, P. Taboada, NiCo2O4-based nanostructured composites for high-performance pseudocapacitor electrodes, Colloids and Surfaces A: Physicochemical and Engineering Aspects. 584 (2020) 124039. doi:10.1016/j.colsurfa.2019.124039.

[30] S.J. Uke, G.N. Chaudhari, A.B. Bodade, S.P. Mardikar, Morphology dependant electrochemical performance of hydrothermally synthesized NiCo2O4 nanomorphs, Materials Science for Energy Technologies. 3 (2020) 289-298.

[31] NiCo2O4-based nanostructured composites for high-performance pseudocapacitor electrodes, Colloids and Surfaces A: Physicochemical and Engineering Aspects. 584 (2020) 124039. doi:10.1016/j.colsurfa.2019. 124039.

[32] S.J. Uke, V.P. Akhare, S.P. Meshram, G.N. Chaudhari, Triethanol amine ethoxylate (TEA-EO) driven controlled synthesis of NiCo2O4 nanostructures, their characterization and supercapacitor performance, Advanced Science, Engineering and Medicine. 10 (2018) 1174-1182.

[33] D. Guragain, C. Zequine, T. Poudel, D. Neupane, R.K. Gupta, S.R. Mishra, Influence of urea on the synthesis of NiCo2O4 nanostructure: morphological and electrochemical studies, Journal of Nanoscience and Nanotechnology. 20 (2020) 2526-2537.

[34] W.D. Wang, P.P. Zhang, S.Q. Gao, B.Q. Wang, X.C. Wang, M. Li, F. Liu, J.P. Cheng, Core-shell nanowires of NiCo2O4@α-Co(OH)2 on Ni foam with enhanced performances for supercapacitors, Journal of Colloid and Interface Science. 579 (2020) 71-81. doi:10.1016/j.jcis.2020.06.048.

[35] H. Yang, C. Zeng, C. Sun, M. Wang, Y. Gao, Mesopores NiCo2O4 nano-needles directly grown on Ni foam as high-performance electrodes for supercapacitors, Materials Letters. 279 (2020) 128523. doi:10.1016/j. matlet.2020.128523.

[36] F. Ren, Z. Tong, S. Tan, J. Yao, L. Pei, A. Abulizi, Ultrathin and porous NiCo2O4 nanosheets based 3D hierarchical electrode materials for high-performance asymmetric supercapacitor, Journal of Electrochemical Energy Conversion and Storage. (2021) 1-11.

[37] Y. Li, Z. Zhang, Y. Chen, H. Chen, Y. Fan, Y. Li, D. Cui, C. Xue, Facile synthesis of a Ni-based NiCo2O4-PANI composite for ultrahigh specific capacitance, Applied Surface Science. 506 (2020) 144646. doi:10.1016/j. apsusc.2019.144646.

[38] Z. Cao, C. Liu, Y. Huang, Y. Gao, Y. Wang, Z. Li, Y. Yan, M. Zhang, Oxygen-vacancy-rich NiCo2O4 nanoneedles electrode with poor crystallinity for high energy density all-solid-state symmetric supercapacitors, Journal of Power Sources. 449 (2020) 227571. doi:10.1016/j.jpowsour.2019.227571.

[39] R. Bai, X. Luo, D. Zhen, C. Ci, J. Zhang, D. Wu, M. Cao, Y. Liu, Facile fabrication of comb-like porous NiCo2O4 nanoneedles on Ni foam as an advanced electrode for high-performance supercapacitor, International Journal of Hydrogen

Energy. 45 (2020) 32343-32354. doi:10.1016/j.ijhydene.2020.08.156.

[40] S.S. Jayaseelan, S. Radhakrishnan, B. Saravanakumar, M.-K. Seo, M.-S. Khil, H.-Y. Kim, B.-S. Kim, Mesoporous 3D NiCo2O4/MWCNT nanocomposite aerogels prepared by a supercritical CO2 drying method for high performance hybrid supercapacitor electrodes, Colloids and Surfaces A: Physicochemical and Engineering Aspects. 538 (2018) 451-459. doi:10.1016/j.colsurfa.2017.11.037.

[41] Preparation of Ordered Mesoporous Carbon/NiCo2O4 Electrode and It...: Ingenta Connect, (n.d.). https://www.ingentaconnect.com/content/apcs/apcs/2012/00000028/00000004/art00016# (accessed April 26, 2021).

[42] C. Young, R. Salunkhe, S. Alshehri, T. Ahamad, Z. Huang, J. Henzie, Y. Yamauchi, High Energy Density Supercapacitors Composed of Nickel Cobalt Oxide Nanosheets on Nanoporous Carbon Nanoarchitectures, Journal of Materials Chemistry A. 5 (2017). doi:10.1039/C7TA01362K.

[43] N.V. Nguyen, T.V. Tran, S.T. Luong, T.M. Pham, K.V. Nguyen, T.D. Vu, H.S. Nguyen, N.V. To, Facile Synthesis of a NiCo2O4 Nanoparticles Mesoporous Carbon Composite as Electrode Materials for Supercapacitor, ChemistrySelect. 5 (2020) 7060-7068.

[44] J. Fang, C. Kang, L. Fu, S. Li, Q. Liu, Fabrication of hollow bamboo-shaped NiCo2O4 with controllable shell morphologies for high performance hybrid supercapacitors, Journal of Alloys and Compounds. 849 (2020) 156317. doi:10.1016/j.jallcom.2020.156317.

[45] P. Siwatch, K. Sharma, N. Singh, N. Manyani, S.K. Tripathi, Enhanced supercapacitive performance of reduced graphene oxide by incorporating NiCo2O4 quantum dots using aqueous electrolyte, Electrochimica Acta. 381 (2021) 138235. doi:10.1016/j.electacta.2021.138235.

[46] L. Hou, W. Yang, X. Xu, B. Deng, J. Tian, S. Wang, F. Yang, Y. Li, In-situ formation of oxygen-vacancy-rich NiCo2O4/nitrogen-deficient graphitic carbon nitride hybrids for high-performance supercapacitors, Electrochimica Acta. 340 (2020) 135996. doi:10.1016/j.electacta.2020.135996.

[47] J. Chen, T. Ma, M. Chen, Z. Peng, Z. Feng, C. Pan, H. Zou, W. Yang, S. Chen, Porous NiCo2O4@Ppy core-shell nanowire arrays covered on carbon cloth for flexible all-solid-state hybrid supercapacitors, Journal of Energy Storage. 32 (2020) 101895. doi:10.1016/j.est.2020.101895.

[48] M. Chatterjee, S. Saha, S. Das, S.K. Pradhan, Advanced asymmetric supercapacitor with NiCo2O4 nanoparticles and nanowires electrodes: A comparative morphological hierarchy, Journal of Alloys and Compounds. 821 (2020) 153503. doi:10.1016/j.jallcom.2019.153503.

[49] P. Salarizadeh, M.B. Askari, M. Seifi, S.M. Rozati, S.S. Eisazadeh, Pristine NiCo2O4 nanorods loaded rGO electrode as a remarkable electrode material for asymmetric supercapacitors, Materials Science in Semiconductor Processing. 114 (2020) 105078. doi:10.1016/j.mssp.2020.105078.

[50] Z. Wei, J. Guo, M. Qu, Z. Guo, H. Zhang, Honeycombed-like nanosheet array composite NiCo2O4/rGO for efficient methanol electrooxidation and supercapacitors, Electrochimica Acta. 362 (2020) 137145. doi:10.1016/j.electacta.2020.137145.

Chapter 8

Effect of Different Metals Doped in Nickel Oxide Nanomaterials on Electrochemical Capacitive Performance

Amar Laxman Jadhav, Sharad Laxman Jadhav and Anamika Vitthal Kadam

Abstract

Recently, the various porous nano metal oxides used for the electrochemical energy storage supercapacitor applications. Some researchers focus on the binary as well as ternary metal oxides and more metal oxide complex composite materials used for the supercapacitors. In the review article focused on the effect of different metals doped in a nickel oxide nano material on the electrochemical capacitive performance, discussion on methodologies, charge storage mechanism, latest research articles and prepared nanostructures. Nowadays nickel oxide is developing electrode material for storage of charge due to its higher thermal stability, excellent chemical stability, cost effective materials, higher theoretical values of specific capacitance, naturally rich and environment friendliness material. The various metals doped in NiO and their composite oxides have shown good structural stability, reversible capacity, long cycling stability and have been also studied nano structured electrode materials for electrochemical supercapacitor applications.

Keywords: Metal oxide, Nickel Oxide, electrochemical supercapacitor, Nanomaterial

1. Introduction

Nowadays, in this research on the rapid growth of electronic portable energy storage devices and hybrids electrical vehicles, the call for high power density and energy density resources has been increased manifold. Supercapacitor also called ultracapacitor or electrochemical capacitors, exhibits higher power density than the normal capacitor and higher energy density than the batteries. The electrochemical capacitor shows faster charge–discharge mechanism behavior and also exhibits long cycle stability. Therefore, supercapacitor or electrochemical capacitor indicates bridge between the normal capacitor and fuel cell, batteries. The electrochemical supercapacitor has two main types based on the charge storage mechanism (i) Electrochemical double layer capacitor (EDLc) is based on electrostatically charge storage mechanism and (ii) pseudocapacitor is electrochemical charge storage mechanism. The carbon-based materials (Activated carbon materials, graphene oxide) used for the preparation of EDLc supercapacitor, transition metal oxide also used for the preparation of active electrode materials for pseudocapacitor and hybrid supercapacitor exhibits

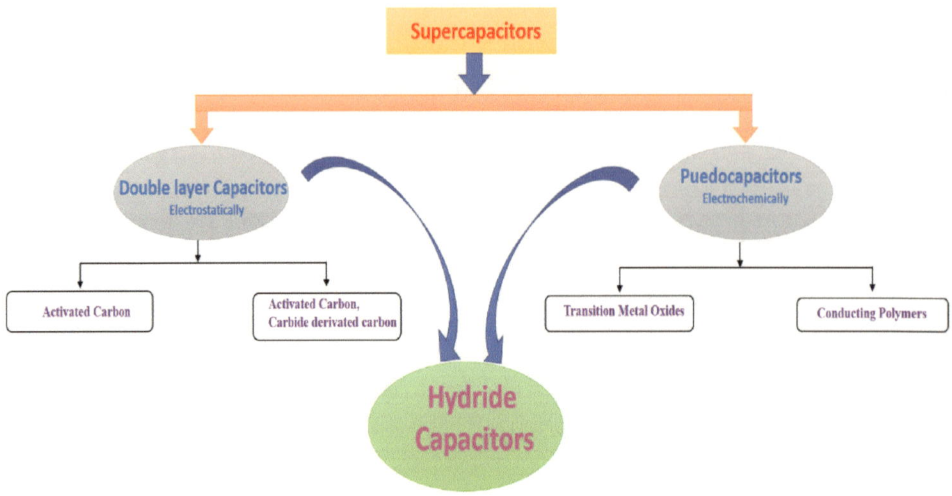

Figure 1.
Classifications of the electrochemical supercapacitor.

intermediate properties between the EDLc and pseudocapacitor behaviors schematic diagram shown **Figure 1**.

The effect of different morphologies on the charge storage at the active electrode materials for the supercapacitor application. The 1D, 2D, and 3D morphology increase the active materials surface area due to an increase in the power density and specific capacitance. It is an important role in electrochemical capacitors. The active electrode materials should be higher specific capacitance, structural stability, and good mechanical to provides long cycling lifetimes. The nano porous active electrode materials prepared by using carbon materials, metal oxides, and conducting polymers such as graphene oxide [1, 2], activated carbon [3] and derivatives of carbide materials [4–10], CuO, NiO, RuO2, Cu2O, Fe2O4, CoO, MnO [11–17] and polyaniline (PANI), polypyrrole (PPy), polythiophene (PTh) [18–22], The transition metal oxide electrode shows excellent properties of electrochemical performance (Specific capacitance, power density, and cycling stability) than the other types of the electrodes.

Generally, research has been carried out on the various transition metal oxide materials like cobalt oxide [23], iridium oxide [24], nickel oxide (NiO) [25], manganese oxide [26], iron Oxide [15], ruthenium oxide [13], and zinc oxide [27]. Currently, research on nickel oxide with other composite electrode materials just like a NiO//graphene oxide, Carbon nanotubes (CNT)//NiO, Ru doped nickel oxide, Cu doped nickel oxide, Cerium doped nickel oxide, and so on [28]. In the fabrications of high-performance supercapacitor at the laboratory, some interruptions occur due to nanostructured morphological structures with the large surface area of the electrode materials. Therefore, the nanostructures of the SEM images are an important parameter for supercapacitor applications. The electrochemical specific capacitance, power density, and cycling stability depend on the morphological structures due to so many researchers work on the synthesis of various types of nanostructured morphologies. But the Nickel oxide electrochemical specific capacitance does not get at maximum (In case of practical its value is 1000F/g) and the theoretical value is 2584F/g. Thus, gets maximum capacitance value, the growth of the nanostructured morphological materials with increased conductivity, lower interfacial resistance, and large surface area of the promising electrode materials is a promising solution.

There are several reports available on the synthesis of nanomaterials and electrochemical characterizing with more conductivity. NiO nanomaterials show excellent physical as well as chemical properties such as mesoporous and hierarchical porous nature, large surface area, and more electronic conductivity. The nanoporous electrode provides a large surface area that can enhance the electrochemical performance because increases the interactions between electrolytes with active electrode materials occurs faradic reaction at the interfacing sites. Further, an available large quantity of porosity nature can be more diffusion of the electrolytic electrons or ions in the electrodes and improve the volume alteration during the charge–discharge cyclic process due to enhances the cycling life of the active electrode materials. The oxide materials can be prepared by the various method can occur different type morphologies such as nanorod, nanoparticles, nanowire, nanoflower, nanotube, nanosheet, nanoneedles by Hydrothermal, Successive ionic adsorption and reaction (SILLAR), chemical precipitation, chemical bath deposition (CBD), sol–gel, solvothermal and electrodeposition methods.

Nickel oxide (NiO) exhibits multiple oxidations states these properties more suitable for the redox reaction or faradic reactions which gets maximum specific capacitance. One of the disadvances is the less electronic conductivity of the electrode. Therefore, large efforts have been dedicated to the manufacture of nanomaterials electrodes the exceptional advantages of some metal oxide doped NiO nanomaterials enhance the higher conductivity. Recently, the literature survey found that the Cu doped NiO nanomaterials, Co-doped nickel oxide, Mn-doped nickel oxide, Cerium doped nickel oxide were found to be very promising for the supercapacitor applications. In this review article, metal oxide doped nickel oxides materials for electrochemical supercapacitor applications have also been discussed briefly. However, to our best knowledge, there is no review found on the development of metal oxides doped NiO nanomaterials electrodes for supercapacitors applications. The energy storage mechanism in metal oxide doped nickel oxide electrode materials also discussed in this review article.

2. Principle of energy storage mechanism in electrochemical supercapacitor

In recent year, the energy storage and energy conversion are a big challenge and concern to the researchers. The excellent electrochemical performance depends on the properties of electrode materials. The electrochemical supercapacitor consists of a three importance parts one is electrode materials and another is an electrolyte, separator. The electrode materials are a most important part in an electrochemical capacitor. In the literature survey, the electrochemical electrical double layer capacitor made by carbon materials as a graphene oxide, activated carbon, carbon nanotubes and derivatives of a carbon materials, pseudocapacitor made by using a metal oxide or conducting polymers and hybrid supercapacitor is a made by using combination of the carbon-based materials and metal oxide. The charge storage mechanism of Electrical double layer capacitor (EDLc) is based on the electrostatically. The electrolytes and electrode materials interfaces on the surface layer in EDLc. The charge storage mechanism of the pseudocapacitor based on the electrochemically i.e. faradic reactions occur in the electrode and electrolytes interface and in the advance hybrid's supercapacitor consist of a both electrostatics and electrochemical charge storage mechanism.

The ideal electrical double layer capacitor electrode materials show rectangular in shape of cyclic charging -discharging cures but if the electrode shows pseudocapacitor behaviors then the curves show the nonlinear rectangular shapes, these nonlinear curves consist of oxidation- reduction peak. This peaks clearly indicates that the electrode materials and electrolytes interfaces occur faradic reactions during the cyclic voltammetry process **Figure 2**.

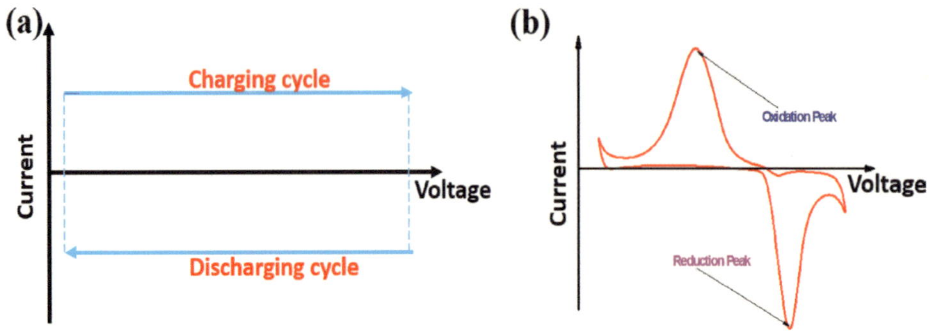

Figure 2.
The charging -discharging cyclic voltammetry curve of electrochemical supercapacitor (a) EDLC curve and (b) pseudocapacitor curve.

2.1 Charge storage mechanism within NiO nano materials

Generally, the metal oxide electrodes show higher power density than the carbon materials and higher electrochemical stability than the conducting polymer material electrodes. The charge store on the surface of the carbon based EDLCs supercapacitors and in puedocapacitor the charge store in porous nano materials, it occurs faradic reactions between the electrode materials and electrolytes. The NiO materials more suitable for supercapacitors applications because they exhibit several required properties.

- NiO exhibit good electrically conductive materials.
- Its shows multi-oxidation states.
- Its shows large surface area of the active materials.
- NiO nano materials shows higher theoretical values of Cs.
- Its exhibit large cycling stability.
- Its shows higher specific capacitance values than the other materials.

NiO have been promising materials for supercapacitor applications due to the exhibits higher theoretical specific capacitance, but in practical case these NiO materials does not get or shows highest specific capacitance values. Sometimes, achieve higher specific capacitance of NiO materials because the NiO shows higher charge storage at the highly porous nanostructure materials, low resistance between electrode and electrolytes, highly conductive substrate. In the practical case NiO based supercapacitors observed in the literature survey, the specific capacitance values 50 to ~1000F/g. The NiO nano materials shows pseudocapacitive nature, during the cyclic voltammetry process these nano-materials exhibits redox reaction mechanism and it's converted to NiOOH and reversible state. Sometimes, during the redox cycles, the cathodic current peaks and anodic current peaks shifted more towards the positive and negative axes as the scan rate was increased. The shifting of peaks currents which is maybe due to the highly accessible surface area of the porous NiO nanostructures and the fast ionic/electronic diffusion rate during redox reaction.

NiO and its binary as well as ternary composite materials prepared as various techniques by using some additives and binder free method. Therefore, the prepared NiO based binary as well as ternary composite or metal doped NiO materials shows lower conductivity and higher interfacing resistance. Due to metal oxides directly synthesized on conductive electrode substrate have the advantage as it can not only result in higher capacitance but also minimize the contact resistance. Various conductive substrates like stainless steel (SS), ITO glass, FTO glass, carbon cloth, carbon mesh, Ni foam, and copper strip electrodes have been used to preparations of the nanostructured metal oxides. It should be mentioned that the electrode metal oxide/hydroxides material deposited directly on to the current collector electrodes is the best choice to minimize the resistance and enhances the electrochemical performance of the electrode materials.

3. Synthesis techniques

Different techniques are used for the synthesis of the nanomaterials. The synthesized nano materials have been formed various nanostructures and proper required porous materials. It is well known that the nano structure morphology of the synthesized electrodes plays an important role in electrochemical redox reactions. The nanostructured electrodes increase the interactions between the electrode and electrolytes due to enhances the performance of the electrochemical performance.

In the chapter, we place advancing a comprehensive summary of the synthesis techniques, pure NiO and metal doped NiO based nanostructures with electrochemical analysis. Following methods briefly discussed one by one.

3.1 Hydrothermal method

Hydrothermal synthesis method is one of the most common used for one-pot synthesis techniques to prepare a wide range of metal oxides [29, 30]. Hydrothermal synthesis process is solution based. We know that these hydrothermal methods provide good crystallinity structures and highly nano porous morphological shape selectivity to oxides-based materials [31]. Generally, the precursors of metal oxides are formed by mixture of reaction ingredients being heated at sealed Teflon-lined stainless-steel autoclave. Solvents under the room temperature to high temperature range and various pressure ranges are used to formations of the nanostructure materials. Generally, the temperature is used higher than 1000C and a pressure will be established mechanically in a closed autoclave system. In hydrothermal techniques to the reaction temperature other parameter like volume of solvent, reaction time also have importance impact on the final synthesis morphology of the materials. By controlling the various parameters such as reaction time, pH value of percussor solutions, reaction temperature and concentration of the precursor solution, it can produce various dimensional (0D, 1D, 2D and 3D) morphologies with large surface area based porous nanostructures. A lot of researcher groups have made optimize the reaction conditions and prepares superior morphologies to enhance the electrochemical performance of the nano materials, which will be discussed in details later section.

3.2 SILAR method

These SILAR methods is a solution-based techniques, it is widely used for the synthesis of the various type of metal oxide/hydroxide thin films, these techniques more commonly used to preparations of the different type nanostructured materials. SILLAR is a simple, cost effective, binder free method and it is appropriate method for synthesis of large-scale of nanostructure materials [32].

3.3 Chemical bath deposition method

The chemical bath deposition (CBD) techniques were first discovered by Nagayama in 1988 [33]. This technique mostly used in prepare of the metal oxide thin films for various applications. The principle of CBD method deals with the immersion of a substrate in a precursor solution [34, 35]. Then the grown-on metal oxide/hydroxide precipitates on the substrate surface to produces a thin well adherent and binder less film. These techniques beneficial due to its low cost, low temperature, binder less, and various adjustment parameter for preparations of different nanostructured materials and also more suitable for large-scale deposition particularly for the preparation of uniform oxide thin films on samples. This method usually needs a strong chemical oxidant or reducing agent to drive reactions to take place. Recently, this technique is very popular in today's for preparations of different types metal oxides as well as hydroxides like NiO thin film nanostructures.

3.4 Electrodeposition method

The electrodeposition is a one of the most widely used method to formations of different metal oxides as well as hydroxides nano materials [36]. This electrodeposition method is based on electrochemical oxidation- reduction (redox) reaction and the metal oxides/hydroxides are deposited on to the conducting substrate electrode. In these techniques, the optimal of the anion and adjustment of appropriate pH of the solution during the deposition is very crucial parameter. Although, this is a simple method to preparations of metal oxides/hydroxides with uniform grown morphology on electrode.

3.5 Spray pyrolysis method

Among the chemical techniques mentioned above, spray pyrolysis is most popular today for the large area thin film formation. Spray solution results directly into oxide formation. It has number of advantages. (1) Doping is easy as required amount of dopant can be added by mixing proper amount of solution of the dopant. (2) Like vapor deposition technique, spray pyrolysis does not require high quality target or vacuum at any stage hence this is one of the great advantages of this technique in the industrial applications. (3) Deposition rate and the thickness of the film can be easily controlled by controlling spray parameters. (4) Deposition in the moderate temperature range 150°C - 500°C is possible. (5) There is no restriction on the size and surface morphology of the substrate. (6) It is possible to prepare multilayer or multi compositional films. Due to these advantages, numbers of conducting and semiconducting materials were prepared by spray pyrolysis technique [37, 38]. In the present work, spray pyrolysis set up (Labotronics make) was used to prepare thin films of cobalt oxide, manganese oxide, manganese doped cobalt oxide, ruthenium oxide, ruthenium doped cobalt oxide and ruthenium doped manganese: cobalt oxide (ternary oxide) thin films by both aqueous as well as non-aqueous routes.

3.6 Microwave-assisted method

Although, the hydrothermal method and solvothermal synthesis method are the most useful methods to preparations of the different types nanostructures materials with controllable structure, size and morphology [39, 40], In hydrothermal method synthesis nanomaterials required more time for the reactions. Therefore, the microwave-assisted method is being used widely for the synthesis of different types nanostructure materials in few minutes. Microwave synthesis has become a popular method. Which substantially reduces the reaction time. The microwave-assisted method can suppress side reactions and provide rapid kinetics of crystallization growth. Using the microwave- solvothermal coupled method; one can not only effectively reduce the reaction time but can also control the morphology. It can produce narrow particle size distribution with high purity and large surface area of the active electrode materials. Therefore, microwave-solvothermal technique is an effective technique to fabricate the different types metal oxide and hydroxides with desired morphology.

4. Pure NiO electrode for supercapacitor applications

The different method used to preparation of the different type's nanostructure nickel oxides/hydroxides. There have been many reports on NiO nanostructures including porous nano/microspheres [41, 42], nanosheets [43], nanoflowers [44] and nanofibers [45]. In general, the proper porous nanostructure plays an important role in electrochemical charge storage mechanism due excellent electrical conductivity and the large surface area [46–48]. The overview of pure NiO prepared by using different synthesis methods and their electrochemical performance is tabulated in **Table 1** [49–63]. The following section briefly discusses on various synthetic routes for the fabrication of pure NiO nanostructures and their supercapacitor properties.

4.1 Hydrothermal method

In the hydrothermal synthesis route provides by the 3D nanostructures of materials with large surface area, these properties more suitable for the supercapacitor applications. Generally, 3D nanostructured electrodes are prepared on the conducting substrate foams like nickel foam. Ni foams exhibits highly porous structures and conductive substrate for synthesis of NiO. In the hydrothermal method's various adjustable parameters like reaction temperature, reaction time, concentration of percussor solutions, pH and so on are used to preparation of the different type's nanostructures. This is due to the increases in the conductivity of the electrodes of the pure NiO. The nanosheets and flower-like morphology of NiO composed of flabbergasted lotus-root- like nanosheets were also fabricated by hydrothermal method [44]. The concentrations of reaction reagent in the reaction medium also have the key role to control the morphology of metal hydroxides and metal oxides.

4.2 SILLAR and chemical bath deposition

This method compared to hydrothermal method, relatively a smaller number of researchers work on preparations of the pure NiO. The specific capacitance SC of NiO were observed in aqueous NaOH and KOH electrolytes observed 129.5 F/g and 69.8 F/g respectively [34]. Xia et al. [57] successfully prepared the NiO monolayer hollow-sphere composed of porous net-like NiO nanoflakes film (SSA 325 m_2/g) by chemical bath deposition using polystyrene sphere template. The SC value for this porous NiO films was found to be 311 F/g with an excellent capacitance retention which is due to the porous structure that could alleviate the structure distortion caused by volume expansion during the cycling process [56].

Nano materials Synthesis techniques	Morphological view	Electrolytes	Specific capacitance	Specific surface area (m^2/g)	Cycling stability	Reference
Hydrothermal	Nanoparticles	6 M KOH	609 F/g at 5 A/g	58.5	500	[41]
Hydrothermal	Nanowires	1 M LiPF4	348 F/g at 10 mV/s	85.18	100	[49]
Hydrothermal	Nanoflakes	2 M KOH	137.7 F/g at 0.2 A/g	107.5	1000	[50]
Hydrothermal	Double-shelled hollow nanospheres	—	612.5 F/g at 0.5 A/g	92.99	1000	[51]
Hydrothermal	Nanocolumns	1 mol/L KOH	390 F/g at 5 A/g	102.4	1000	[52]
Hydrothermal	Nanosheets-assembles	2 M KOH	989 F/g at 3 mV/s	—	1000	[53]
Hydrothermal	Pine-cones	2 M KOH	337 F/g at 2 mV/s	265	100	[54]
Hydrothermal	Nanoflakes	2 M KOH	411 F/g at 0.2 A/g	227	100	[55]
Chemical bath deposition	Monolayer porous hollow-sphere arrays	—	311 at 1 A/g	325	—	[56]
Chemical bath deposition	Flackes	1MNaOH	129.5F/g	—	—	[34]
SILLAR	nanoflakes	2 M KOH	674F/g	122.36	2000	[25]
SILLAR	nanosheets	a 1-(2,3 -dihydroxypropyl)-3-methylimidazolium hydroxide	205.5	36.3	5000	[57]
Electrodeposition	Core/shell	—	1635 F/g at 2 mV/s	—	—	[58]
Electrodeposition	Nanoporous film	—	1776 F/g at 1 mV/s	264	—	[59]
Electrodeposition	nanoflakes	1MKOH	222F/g			[60]
Electrodeposition	Porous NiO	1MKOH	351F/g	—	—	[61]
Microwave	Flower-like hollow nanospheres	2 M KOH	585 F/g at 5 A/g	176	1000	[39]
Microwave	Nanoplatelets	2MNaOH	1200 F/g at 1 A/g	e	1000	[40]

Nano materials Synthesis techniques	Morphological view	Electrolytes	Specific capacitance	Specific surface area (m^2/g)	Cycling stability	Reference
Microwave	Hierarchical porous ball-like surface	Hydroxide Ions	420 F/g at 0.5 A/g	125	1600	[55]
Spray method	larger grains	1MKOH	23 mF/cm2			[62]
Spray method	heaps	2MKOH	405F/g	—	1000	[64]
Spray method	Small pores	2MKOH	564 F/g	—	1000	[63]

Table 1.
Pure NiO prepared using various methods and their electrochemical supercapacitor performances.

4.3 Electrodeposition method

In 1997 year, the porous NiO nanostructure materials were reported by Srinivasan et al. [65], where shows a very little capacitance value is 59 F/g. However, in 2004 year, the capacitance value observed 138 F/g has been reported for 3D NiO on stainless steel conducting substrate [66]. 1D mesoporous core shell structure shows the hexagonal lyotropic $Ni(OH)_2$ synthesized by electrodeposition [67]. The main drawback of NiO is a lower electrical conductivity, and the achieving higher conductivity has a great, the NiO/ITO showed increased conductivity and thus improved the SC (1025 F/g) compared to NiO/Ti (416 F/g) [68].

4.4 Spray pyrolysis method

Among numerous chemical techniques mentioned in schematic (**Figure 3**) SPM is the most popular today because of its applicability to produce variety of doped and undoped metal oxide films [69]. The basic principle involved in SPM is the pyrolytic decomposition of salts of a desired compound onto the preheated substrates. The atomization of the spray solution into a spray of fine droplets also depends on the geometry of the spraying nozzle and pressure of a carrier gas. Every sprayed droplet reaching the surface of the hot substrate undergoes pyrolytic (endothermic) decomposition and forms a single crystallite or clusters of crystallites as a product. The remaining volatile byproducts and solvents escape out in the form of vapor phase. The substrates provide thermal energy for the decomposition and subsequent recombination of the constituent species, followed by sintering and crystallization of the clusters of crystallites and thus coherent films are formed. The required thermal energy is different for the different materials and the solvents used.

Nickel oxide (NiO) films can be prepared using various chemical methods. Among these, spray method is a mechanically simple, cost-effective, and large

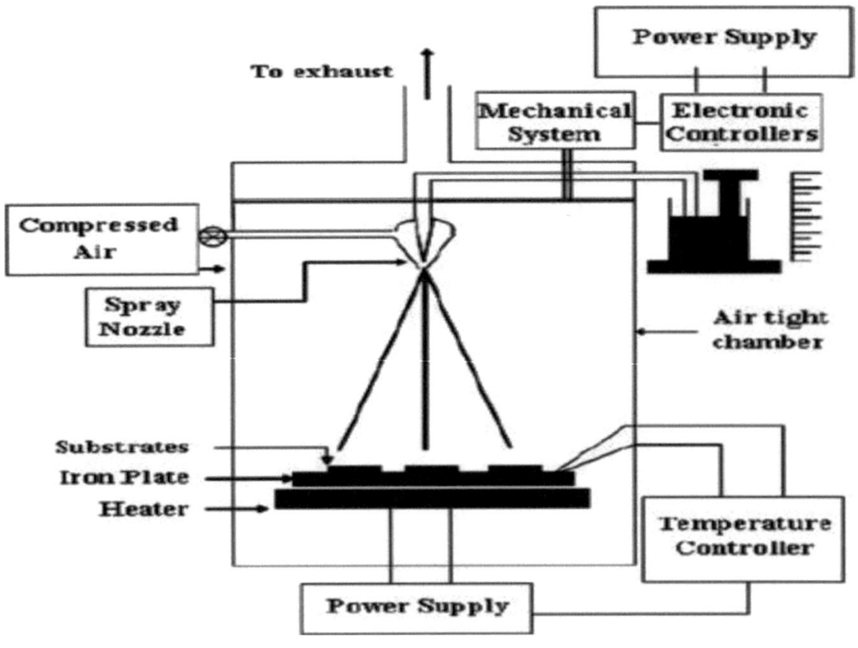

Figure 3.
Schematic spray pyrolysis deposition method. (Adapted with permission from Ref. [69]. Copyright 2021, Elsevier.)

surface deposition method. The various precursors are used for preparation of NiO thin films electrodes using different in gradient sources like nickel nitrate, nickel acetate and nickel chloride [70]. Yadav et.al reported specific capacitance of the pure NiO is 564F/g at 1A/g in 2 M KOH electrolytes with 1000 cyclic stability [70]. Kate et.al reported that the NiO thin films were successfully deposited using spray method, the observed specific capacitance values is 1000F/g at 5 mV/s scan rate [71].

5. Metals doped in NiO nanomaterials for supercapacitor applications

A great development has been achieved in developing low cost, higher conductivity, porous materials and more simple methods for the synthesis of various metal doped oxide electrodes for electrochemical supercapacitor applications. In the previous point discussed, pure NiO has drawn rigorous research interests due to its promising properties and some drawback of the electrochemical analysis. However, pure NiO exhibits lower specific capacitance values (SC) and it is providing lower electrochemical stability. There are several reasons for the low stability and low capacitance of pure NiO. Such as conductivity of NiO materials is very poor, does not proper electrolytes interactions of the nanomaterials and so on. But if we want to increase the electrochemical stability with capacitance then we have to dope the proper metals, so that the conductivity of the nanomaterials will increases and also the capacitance will be increased. In the following **Table 2** [28, 72–80] shows metal oxide doped in NiO and its effect on electrochemical performance. Various method is used to formations of the metal doped NiO nanostructures like hydrothermal, spray, sol–gel, chemical bath, facile chemical synthesis and so on.

In the hydrothermal method various adjustable parameters used for the for the synthesis of different types nanostructured morphologies. Such as temperature controlled, concentration of the percussor, reaction time and different types

Different Metal Doped NiO	Nano materials Synthesis techniques	Morphological view	Electrolytes	Specific capacitance	Cycling stability	Reference
Mn-NiO	Facile chemical synthesis route	Nanoparticle	6MKOH	369.6F/g		[72]
Mn-NiO	Hydrothermal	Nanostructured arrey	6MKOH	1166f/g	5000	[73]
Cu-NiO	citrate-gel	Particle size		559F/g	50	[74]
Al-NiO	Hydrothermal	Nanosheet	1MKOH	2253F/g	5000	[75]
Co-NiO	Spray method	Grannular	2MKOH	835F/g	1000	[76]
Co-NiO	Laser deposition	Flowerlike	1MKOH	720F/g	1000	[77]
Ce-NiO	Chemical method	Porous layer	1 M LiOH, NaOH and KOH	1500F/g	2000	[78]
Ce-NiO	Sol–gel	Nanoflaske	1MKOH	2424F/g	2000	[28]
Ce-NiO	Sol–gel	Spongy like		110F/g	1000	[79]
Cu-Co-Ce-Ni	Hydrothermal	Nanoflaske based nanoflower	1MKOH	2696F/g	3000	[80]

Table 2.
Different Metal doped in NiO prepared using various methods and their structural, electrochemical supercapacitor performances.

substrate foams are used for the preparations of the large surface area of the nanostructured morphologies. In schematic **Figure 4** shows effect of all above parameters on the nanomaterials and formations of the different nano structures [81]. The 1D or 2D structures like NiO nanorod, $Ni(OH)_2$ nano-wall and Co_3O_4 nanowire/nanosheet arrays can be attained by using simple hydrothermal of directly putting the substrates into precursor solutions (contain metal salts and alkali), maintaining for a certain time at appropriate temperature and following annealing treatment (**Figure 4**). The morphological structures and their size strongly depend only on the reaction conditions, such as reaction temperature, reaction time, and concentration of the precursor solutions and its ratios proportionality of the reactants. The observed structures of the NiO are the small nano array and Nickel hydroxides small nano wire like morphological structures.

In sol gel synthesis is a one of the widely used method for the formation of the large number nano materials. Saraynya et.al shows that the cerium doped nickel oxide more active and suitable materials for the supercapacitor applications [28]. They are observed highest specific capacitance value is 2444 F/g at 5 mV/s at a 1%

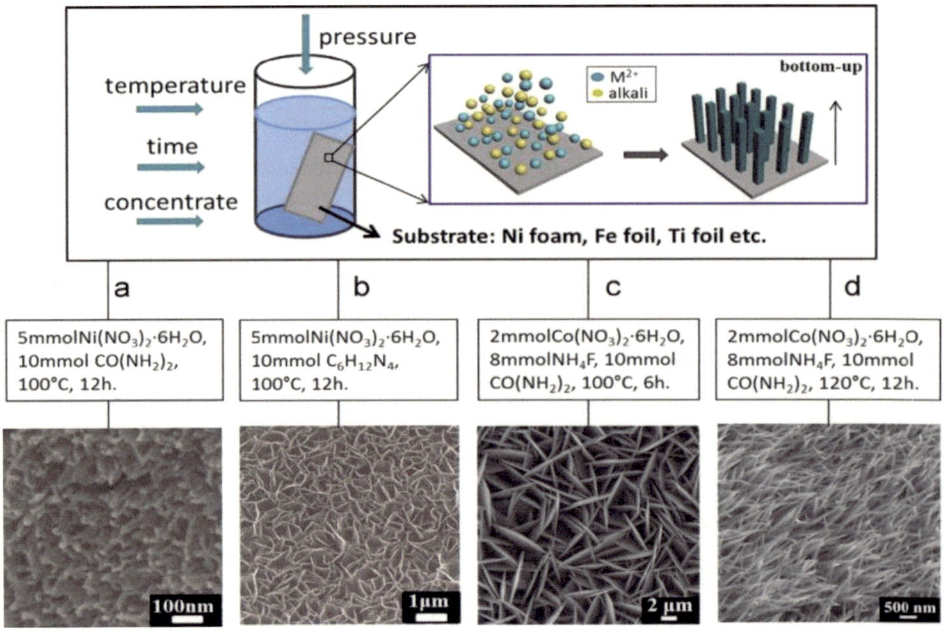

Figure 4.
Schematic hydrothermal deposition with using various parameter. (Adapted with permission from Ref. [81]. copyright 2021, Elsevier).

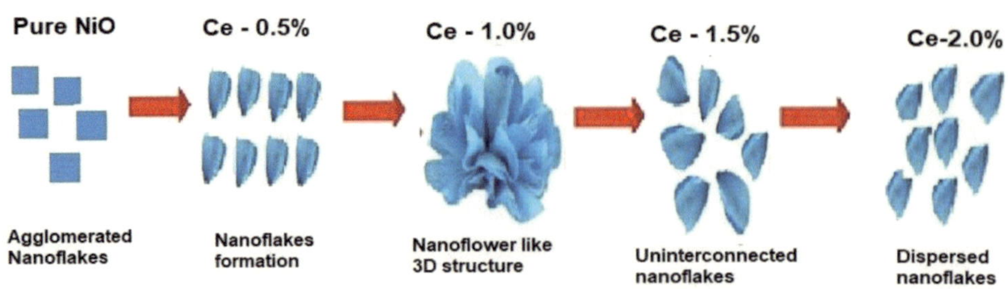

Figure 5.
The effect on nanostructure by using various doping percentage in NiO. (Adapted with permission from Ref. [28]. Copyright 2021, Elsevier).

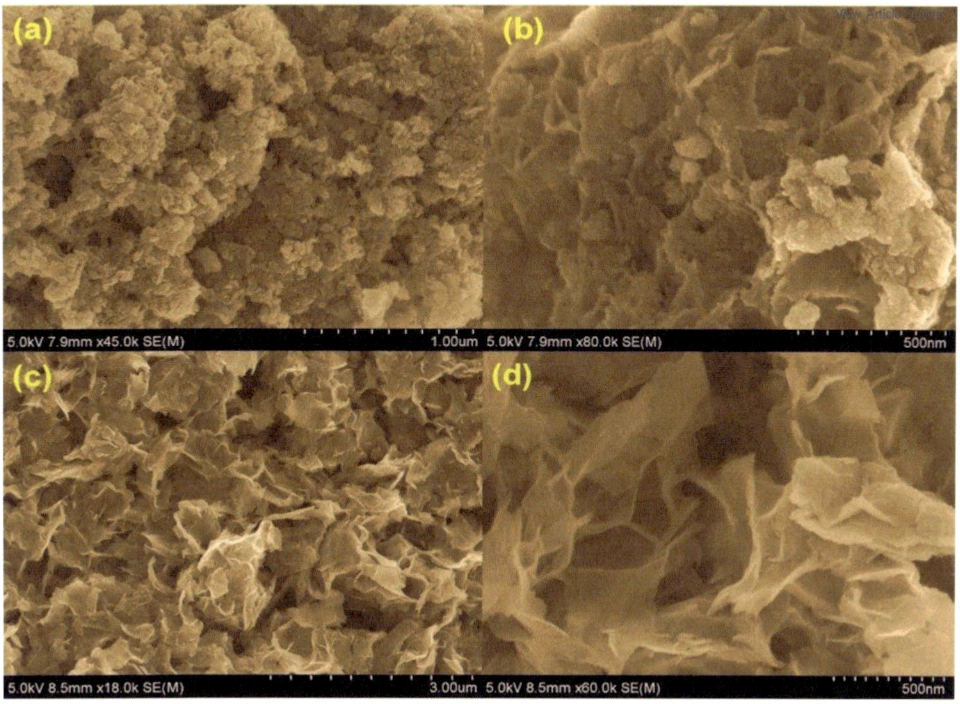

Figure 6.
(a and b) The SEM images of pure NiO and (c and d) The SEM images of 1% Ce doped NiO. (Adapted with permission from Ref. [28]. Copyright 2021, Elsevier).

cerium doped nickel oxide. In a **Figure 5** shows the pure NiO is lower value of specific capacitance because these NiO clearly indicates agglomerated nanostructured morphologies, but in the cerium doped NiO exhibits highest specific capacitance due to the morphologies flake like and large active electrode surface area for the access of the electrons. Here in **Figure 6** shows percentage of the metal dopant increases the structure of the nanomaterials change, it clearly indicates mostly important parameter for the proper formations of the nanostructured morphology for charge storage supercapacitor applications. The 1% Cerium doped NiO formatted 3D nanoflower like structure with accessible large surface [28].

6. Materials selection and challenges of the supercapacitor

The effect of transition metal oxide doped NiO by spray method shows more electrochemical performance than the pure NiO. Kate et.al show that the cobalt doped nickel oxide highest specific capacitance value is 835F/g at 5 mV/s in a 2MKOH electrolytes. Sharanya et.al shows that the the cerium doped nickel oxide is exhibits pseudocapacitive nature and good candidate materials for the supercapacitor applications, his shows that the highest specific capacitance of the Ce doped NiO is 2444F/g [28].

6.1 Selection of electrode materials for supercapacitor applications

Electrochemical energy storage devices play an important role in developing green energy for future to the society. After evaluating the published literature survey, we noticed that a great research has been carried out on electrochemical supercapacitor applications. A great challenge on the to investigate the electrochemical highly performed electrode materials. The various researches focused on the preparation of the large surface of the material morphologies and enhancing conductivity, and electrolytes to obtain high energy and power densities with long cycle life. Therefore, it is necessary to the selection of electrode materials for the electrochemical supercapacitor. Therefore, herein we proposed some designing high-performance electrode materials for supercapacitors, such as specific surface area, proper selecting electrolytes, conductivity of the electrode's materials and design more porous different types nanostructures.

7. Challenges in the supercapacitor applications

Various challenges in the electrochemical supercapacitor such as enhances specific surface area of the electrode materials, enhances the conductivity of the electrode materials, maintain proper thickness of the electrodes, proper electrolytes used in charge storage mechanism, fabrications of the device, suitable separator used between the electrodes, leakage problem in devices and maintain the equivalent series resistance of the electrode materials. All of the above challenges most important in the supercapacitor application.

8. Conclusions

In the chapter, we have scientifically drawn the recent progress on a transition metal doped nickel oxide (NiO) as the energy storage materials for supercapacitor applications. The effect of the metal doped nickel oxide on the supercapacitor and developing nanostructures of pure NiO and metal oxide doped NiO based pseudocapacitor electrodes have been discussed. If you want to get maximum capacitance, you need to have a specific surface area, it is crucial parameter to obtain suitable morphologies of electrode materials. Clearly indicates that the various nanostructures of the electrodes such as flower, flake, nanobelt, nanowire, nanorod, hollow, core-shell, granular particles thin films are needed to improve the electrochemical performances further. The specific surface area and conductivity of the electrodes are two most important critical parameters that determine the supercapacitor performance which has to be optimized.

Nickel oxide is a semiconductor material, it shows lower electric conductivity so it has the same effect on the electron motion and hence the effect on the specific capacitance of the supercapacitor. To improve electric conductivity, NiO is often combined with nanostructured conductive transition metal oxides such as cerium, copper, aluminum, and magnesium to produces Metal doped NiO based electrodes. In this way, good electrical conductivity and rich electroactive sites for the electrolyte ions are obtained. The metal dope nickel oxide is shows pseudocapacitive nature. The doping of other metal oxides can also introduce impurity band effects and can enhance the electrochemical performance.

It is critical for researchers to improve both synthesis conditions and material qualities in order to fully leverage the potential of NiO-based electrode materials. High specific capacitance and long-term cycle stability are also concerning that must be addressed. This is the focus of the authors' on-going work. Other hand the synthesis methodologies, there are many issues pertaining to the measurement techniques and electrode preparation process that require attention. Furthermore, engineering factors like fabrication of electrodes, choice of electrolytes, membrane separators and packaging are not well established and thus need extensive investigation.

Conflict of interest

The authors declare no conflict of interest.

Author details

Amar Laxman Jadhav*, Sharad Laxman Jadhav and Anamika Vitthal Kadam
Lab of Electrochemical Energy Studies, Department of Physics, The Institute of Science, Mumbai (Dr. Homi Bhabha State University), Mumbai, India

*Address all correspondence to: jadhavamar50@yahoo.com

References

[1] A.G. Pandolfo, A.F. Hollenkamp, Carbon properties and their role in supercapacitors, J. Power Sources 157 (2006) 11-27. DOI: 10.1016/j.jpowsour.2006.02.065

[2] Reyhaneh Fazel Zarandi, Behzad Rezaei, Hassan S. Ghaziaskar, Ali Asghar Ensaf, Synthesis of graphene oxide-polychrysoidine nanocomposite for supercapacitor applications, Journal of Energy Storage, (2020), 29, 101334. https://doi.org/10.1016/j.est.2020.101334

[3] Sultan Ahmed, AhsanAhmed, M.Rafata, Supercapacitor performance of activated carbon derived from rotten carrot in aqueous, organic and ionic liquid-based electrolytes, Journal of Saudi Chemical Society,(2018),22, 8, 993-1002. doi.org/10.1016/j.jscs.2018.03.002

[4] K. Leitner, A. Lerf, M. Winter, J. Besenhard, S. Villar-Rodil, F. Suarez-Garcia,A. Martinez-Alonso, J. Tascon, Nomex-derived activated carbon fibers as electrode materials in carbon based supercapacitors, J. Power Sources 153 (2006) 419-423. DOI:10.1016/j.jpowsour.2005.05.078

[5] C. Kim, Y.-O. Choi, W.J. Lee, K.S. Yang, Characteristics of supercapaitor electrodes of PBI-based carbon nanofiber web prepared by electrospinning, Electrochimica Acta 50 (2004) 883-887. https://doi.org/10.1016/j.electacta.2004.02.071

[6] D.R. Dreyer, S. Park, C.W. Bielawski, R.S. Ruoff, The chemistry of graphene oxide, Chem. Soc. Rev.,9 (2010) 228-240, https://doi.org/10.1039/B917103G

[7] C. Gomez-Navarro, J.C. Meyer, R.S. Sundaram, A. Chuvilin, S. Kurasch,M. Burghard, K. Kern, U.Kaiser, Atomic Structure of Reduced Graphene Oxide, Nano Lett., 10 (2010) 1144-1148, doi.org/10.1021/nl9031617

[8] A. Lerf, H. He, M. Forster, J. Klinowski, Synthesis of Graphene Oxide (GO) by Modified Hummers Method and Its Thermal Reduction to Obtain Reduced Graphene Oxide (rGO), J. Phys. Chem. B ,102,(1998) 4477-4482, https://doi.org/10.1021/jp9731821

[9] M.M. Sk, C.Y. Yue, Layer-by-layer (LBL) assembly of graphene with p- phenylenediamine (PPD) spacer for high performance supercapacitor applications, RSC Adv. 4 (2014) 19908-19915, https://doi.org/10.1039/C4RA02652G

[10] E. Frackowiak, K. Metenier, V. Bertagna, F. Beguin, A Single-Step Process for Preparing Supercapacitor Electrodes from Carbon Nanotubes, Appl. Phys. Lett. 77 (2000) 2421-2423, doi:10.1063/1.1290146

[11] SV Kambale, AL Jadhav, RM Kore, AV Thakur, BJ Lokhande, Cyclic Voltammetric Study of CuO Thin Film Electrodes Prepared by Automatic Spray Pyrolysis, Macromolecular Symposia,(2019), 387 (1), 1800213, https://doi.org/10.1002/masy.201800213

[12] A.D. Jagadale, V.S. Kumbhar, D.S. Dhawale, C.D. Lokhande, Potentiodynamically deposited nickel oxide (NiO) nanoflakes for pseudocapacitors, Journal of Electroanalytical Chemistry, (2013), 704, 90-95, https://doi.org/10.1016/j.jelechem.2013.06.020

[13] V.Subramanian, Sean C.Hall, Patricia H.Smithb, Rambabu, Mesoporous anhydrous RuO2 as a supercapacitor electrode material, Solid State Ionics ,(2004),175,1-4,511-515, doi.org/10.1016/j.ssi.2004.01.070

[14] Rudra Kumar, Prabhakar Rai,Ashutosh Sharma, Facile synthesis of Cu2O microstructures and their

morphology dependent electrochemical supercapacitor properties, RSC Adv., (2016),6, 3815-3822, https://doi.org/10.1039/C5RA20331G

[15] Huailin Fan, Ruiting Niu, Jiaqi Duan, Wei Liu, Wenzhong Shen, Fe3O4@Carbon Nanosheets for All-Solid-State Supercapacitor Electrodes, ACS Appl. Mater. Interfaces (2016), 8, 30, 19475– 19483, https://doi.org/10.1021/acsami.6b05415.

[16] Xuezhi Sun, Yongxin Lu, Tongtong Li, Shuaishuai Zhao, Zhida Gao, Yan- Yan Song, Metallic CoO/Co heterostructures stabilized in an ultrathin amorphous carbon shell for high-performance electrochemical supercapacitive behaviour, J. Mater. Chem. A, (2019),7, 372-380, https://doi.org/10.1039/C8TA09733J

[17] Ming Huang,Fei Li, Fan Dong,Yu Xin Zhang,Li Zhang, MnO2-based nanostructures for high-performance supercapacitors, J. Mater. Chem. A, (2015),3, 21380-21423,https://doi.org/10.1039/C5TA05523G

[18] M.M. Sk, C.Y. Yue, Synthesis of polyaniline nanotubes using the self-assembly behavior of vitamin C: a mechanistic study and application in electrochemical supercapacitors, J. Mater. Chem. A 2 (2014) 2830-2838, https://doi.org/10.1039/C3TA14309K

[19] Y. Shi, *L. Pan*, B. Liu, Y. Wang, Y. Cui, Z. Bao, G. Yu, Nanostructured conductive polypyrrole hydrogels as high-performance, flexible supercapacitor electrodes J. Mater. Chem. A 2 (2014) 6086-6091, https://doi.org/10.1039/C4TA00484A

[20] B. Muthulakshmi, D. Kalpana, S. Pitchumani, N. Renganathan, Electrochemical deposition of polypyrrole for symmetric supercapacitors, J. Power Sources 158 (2006),1533-1537, DOI:10.1016/j.jpowsour.2005.10.013

[21] A. Laforgue, P. Simon, C. Sarrazin, J.-F. Fauvarque, Polythiophene-based supercapacitors,J. Power Sources 80 (1999) 142-148, https://doi.org/10.1016/S0378-7753(98)00258-4

[22] B. Senthilkumar, P. Thenamirtham, R.K. Selvan, Structural and electrochemical properties of polythiophene, Appl. Surf. Sci. 257 (2011) 9063-9067, doi.org/10.1016/j.apsusc.2011.05.100

[23] Arshid Numan,Navaneethan Duraisamy, Fatin Saiha Omar,Y. K. Mahipal,K. Ramesha, S. Ramesh, Enhanced electrochemical performance of cobalt oxide nanocube intercalated reduced graphene oxide for supercapacitor application, RSC Adv., (2016),6, 34894-34902, https://doi.org/10.1039/C6RA00160B

[24] D.-Q. Liu, S.-H. Yu, S.-W. Son, S.-K. Joo, Electrochemical Performance of Iridium Oxide Thin Film for Supercapacitor Prepared by Radio Frequency Magnetron Sputtering Method, ECS Trans, 16,(2008) 103- 109, https://iopscience.iop.org/article/10.1149/1.2985632/meta

[25] Girish S. Gund, C. D. Lokhande, Ho Seok Park,Controlled Synthesis of Hierarchical Nanoflake Structure of NiO Thin Film for Supercapacitor Application, Journal of Alloys and Compounds,(2018),741, 549-556, doi: 10.1016/j.jallcom.2018.01.166.

[26] Trung Dung Dang, Thi Thu Hang Le, Thi Bich Thuy Hoang,Thanh Tung Mai, Synthesis of nanostructured manganese oxides-based materials and application for supercapacitor, Adv. Nat. Sci: Nanosci. Nanotechnol,(2015), 6, 025011. https://iopscience.iop.org/article/10.1088/2043-6262/6/2/025011

[27] Xuechun Xiao, Bingqian Han, Gang Chen, Lihong Wang ,Yude Wang, Preparation and electrochemical performances of carbon sphere@ZnO

core-shell nanocomposites for supercapacitor applications, Sci. Rep. (2017) 7, 40167; doi: 10.1038/srep40167.

[28] P. E. Saranya,S. Selladurai, Mesoporous 3D network Ce-doped NiO nanoflakes as high performance electrodes for supercapacitor applications, New J. Chem., 2019, DOI:10.1039/C9NJ00097F.

[29] Y. Cui, C. Wang, S. Wu, G. Liu, F. Zhang, T. Wang, Lotus-root-like NiO nanosheets and flower-like NiO microspheres: synthesis and magnetic properties, Cryst Eng.Comm, 13 (2011) 4930-4934,https://doi.org/10.1039/C1CE05389B

[30] M. Liu, J. Chang, J. Sun, L. Gao, A facile preparation of NiO/Ni composites as high-performance pseudocapacitor materials, RSC Adv. , (2013) 8003-8008, https://doi.org/10.1039/C3RA23286G

[31] F. Jiao, A.H. Hill, A. Harrison, A. Berko, A.V. Chadwick, P.G. Bruce, Synthesis of Ordered Mesoporous NiO with Crystalline Walls and a Bimodal Pore Size Distribution, J. Am. Chem. Soc. 130 (2008) 5262- 5266, https://doi.org/10.1021/ja710849r

[32] Yunus Akaltuna, Tuba Çayır, Fabrication and characterization of NiO thin films prepared by SILAR Method, Journal of Alloys and Compounds, (2015),625, 144- 148, https://doi.org/10.1016/j.jallcom.2014.10.194.

[33] H. Nagayama, H. Honda, H. Kawahara, A New Process for Silica Coating, J. Electrochem. Soc. 135, (1988),2013-2016, https://iopscience.iop.org/article/10.1149/1.2096198

[34] A.I. Inamdar, Y. Kim, S.M. Pawar, J.H. Kim, H. Im, H. Kim, chemically grown, porous, nickel oxide thin-film for electrochemical supercapacitors, J. Power Sources, 196, (2011) ,2393- 2397, https://doi.org/10.1016/j.jpowsour.2010.09.052

[35] Dhananjay Mugle, Ghanshyam Jadhav, Short review on chemical bath deposition of thin film and characterization, AIP Conference Proceedings 1728, 020597 (2016); https://doi.org/10.1063/1.4946648

[36] TS Ghadge, AL Jadhav, YM Uplane, AV Thakur, SV Kamble, BJ Lokhande, Controlled synthesis, structural, morphological and electrochemical study of Cu (OH) 2@ Cu flexible thin film electrodes prepared via aqueous–non-aqueous Routes, Journal of Materials Science: Materials in Electronics 32 (7),(2021), 9018-903, https://doi.org/10.1007/s10854-021-05572-8

[37] A.L. Jadhav , S.V. Kambale , R.M. Kore , B.J. Lokhande, Capacitive Study of Nickel Oxide Thin Films Prepared by Spray Pyrolysis, International Journal of Fracture and Damage Mechanics,5,2,(2019), https://www.researchgate.net/profile/Al- Jadhav-2/publication/349053316

[38] C.D.Lokhande, S.H. Pawar, Effect of aluminum doping on the properties of PEC cells formed with CdS:Al films, Material Research Bulletin, 18 ,(1983), 21, https://doi.org/10.1016/0025-5408(83)90035-1

[39] C.Y. Cao, W. Guo, Z.-M. Cui, W.-G. Song, W. Cai, Microwave-assisted gas/liquid Interfacial synthesis of flowerlike NiO hollow nanosphere precursors and their application as supercapacitor electrodes, J. Mater. Chem., 21,(2011), 3204- 3209, https://doi.org/10.1039/C0JM03749D

[40] M. Khairy, S.A. El-Safty, Mesoporous NiO nanoarchitectures for electrochemical energy storage:influence of size, porosity, and morphology, RSC Adv., 3, (2013), 23801-23809, https://doi.org/10.1039/C3RA44465A

[41] G. Cheng, Y. Yan, R. Chen, From Ni-based nanoprecursors to NiO

nanostructures: morphology-controlled synthesis and structure-dependent electrochemical behavior, New J. Chem, 39, (2015),676-682, https://doi.org/10.1039/C4NJ01398K

[42] C. Yuan, X. Zhang, L. Su, B. Gao, L. Shen, Facile synthesis and self-assembly of Hierarchical porous NiO nano/micro spherical superstructures for high performance supercapacitors, J. Mater. Chem, 19, (2009), 5772-5777, https://doi.org/10.1039/B902221J

[43] J. Zhu, Z. Gui, From layered hydroxide compounds to labyrinth-like NiO and Co3O4 porous nanosheets, Mater. Chem. Phys., 118, (2009), 243-248, doi.org/10.1016/j.matchemphys.2009.07.044

[44] Le Xin Song,Zheng Kun Yang, Yue Teng,Juan Xiaa, Pu Du,Nickel oxide nanoflowers: formation, structure, magnetic property and adsorptive performance towards organic dyes and heavy metal ions, J. Mater. Chem. A, (2013),1, 8731-8736, https://doi.org/10.1039/C3TA12114C

[45] Y. Qiu, J. Yu, X. Zhou, C. Tan, J. Yin, Synthesis of Porous NiO and ZnO Submicro- and Nanofibers from Electrospun Polymer Fiber Templates, Nanoscale Res. Let t., 4 (2009) 173-177, https://doi.org/10.1007/s11671-008-9221-6

[46] P. Simon, Y. Gogotsi, Materials for electrochemical capacitors, Nat. Mater 7 (2008) 845-854, DOI: 10.1038/nmat2297

[47] X. Xia, J. Tu, Y. Zhang, X. Wang, C. Gu, X.-B. Zhao, H.J. Fan, Porous Hydroxide Nanosheets on Preformed Nanowires by Electrodeposition: Branched Nanoarrays for Electrochemical Energy Storage, ACS Nano 6 (2012) 5531-5538, https://doi.org/10.1021/cm302416d

[48] Shuchao Hu, Ling Sun,Mingxing Liu, Hongda Zhu, Huiling Guo, Hongmei Suna ,Honghao Sun, A highly dispersible silica pH nanosensor with expanded measurement ranges, New J. Chem.,(2015),39, 4568-4574, https://doi.org/10.1039/C4NJ02419B

[49] Dawei Su, Hyun-Soo Kim, Woo-Seong Kim, Prof. Dr. Guoxiu Wang,Mesoporous Nickel Oxide Nanowires: Hydrothermal Synthesis, Characterisation and Applications for Lithium-Ion Batteries and Supercapacitors with Superior Performance, Chem. Eur. J. ,(2012), 18, 8224 – 8229, https://doi.org/10.1002/chem.201200086

[50] Yan-zhenZheng,Hai-yangDing,Mi-linZhang, Preparation and electrochemical properties of nickel oxide as a supercapacitor electrode material, Materials Research Bulletin,44,2, (2009),403-407, https://doi.org/10.1016/j.materresbull.2008.05.002

[51] Z. Yang, F. Xu, W. Zhang, Z. Mei, B. Pei, X. Zhu, Controllable preparation of multishelled NiO hollow nanospheres via layer-by-layer self-assembly for supercapacitor application, J. Power Sources, 246, (2014), 24-31, https://doi.org/10.1016/j.jpowsour.2013.07.057

[52] X. Zhang, W. Shi, J. Zhu, W. Zhao, J. Ma, S. Mhaisalkar, *T. Maria*, Y. Yang,H. Zhang, H. Hng, Q. Yan, Synthesis of porous NiO nanocrystals with controllable surface area and their application as supercapacitor electrodes, Nano Res. ,3 , (2010), 643-652, https://doi.org/10.1007/s12274-010-0024-6.

[53] K.K. Purushothaman, I. Manohara Babu, B. Sethuraman, G. Muralidharan, Nanosheet-Assembled NiO Microstructures for High-Performance Supercapacitors, ACS Appl. Mater. Interfaces, 5, (2013), 10767-10773, https://doi.org/10.1021/am402869p

[54] S.K. Meher, P. Justin, G.R. Rao, Pine-cone morphology and

pseudocapacitive behavior of nanoporous nickel oxide, Electrochimica Acta, 55, (2010), 8388- 8396, https://doi.org/10.1016/j.electacta.2010.07.042.

[55] P. Justin, S.K. Meher, G.R. Rao, J. Phys. Chem. C, Tuning of Capacitance Behavior of NiO Using Anionic, Cationic, and Nonionic Surfactants by Hydrothermal Synthesis, 114, (2010), 5203-5210, https://doi.org/10.1021/jp9097155

[56] X.-H. Xia, J.-P. Tu, X.-L. Wang, C.-D. Gu, X.-B. Zhao, Self-supported hydrothermal synthesized hollow Co3O4nanowire arrays with high supercapacitor capacitance, J. Mater. Chem., 2011,21, 9319-9325, https://doi.org/10.1039/C1JM10946D`

[57] S C Bhise, D V Awale, M M Vadiyar, S K Patil, U V Ghorpade, B N Kokare, J H Kim, S S Kolekar, A mesoporous nickel oxide nanosheet as an electrode material for supercapacitor application using the 1-(2 ,3 -dihydroxypropyl)-3-methylimidazolium hydroxide ionic liquid electrolyte, Bull. Mater. Sci, (2019) 42, 263, https://doi.org/10.1007/s12034-019-1961-7

[58] A.K. Singh, D. Sarkar, G.G. Khan, K. Mandal, J. Mater. Chem. A, Unique hydrogenated Ni/NiO core/shell 1D nano-heterostructures with superior electrochemical performance as supercapacitors, 1, (2013), 12759-12767, https://doi.org/10.1039/C3TA12736B

[59] K. Liang, X. Tang, W. Hu, High-performance three-dimensional nanoporous NiO film as a supercapacitor electrode, J. Mater. Chem., 22, (2012), 11062-11067, https://doi.org/10.1039/C2JM31526B

[60] A.D. Jagadale, V.S. Kumbhar, D.S. Dhawale, C.D. Lokhande, Potentiodynamically deposited nickel oxide (NiO) nanoflakes for pseudocapacitors, Journal of Electroanalytical Chemistry, 704,(2013),90-95, http://dx.doi.org/10.1016/j.jelechem.2013.06.020

[61] Mao-Sung Wu*, Yu-An Huang, Chung-Hsien Yang, Jiin-Jiang Jow, Electrodeposition of nanoporous nickel oxide film for electrochemical capacitors, International Journal of Hydrogen Energy, 32 ,(2007) ,4153 – 4159, doi:10.1016/j.ijhydene.2007.06.001

[62] A. S. Devasthali, S. G. Kandalkar, Preparation and characterization of spray deposited nickel Oxide (NiO) thin film electrode for supercapacitor, IOSR Journal of Computer Engineering, e-ISSN:2278-0661, p-ISSN:2278-8727,47-51

[63] Abhijit Yadav , U.Chavan, Influence of substrate temperature on electrochemical supercapacitive performance of spray deposited nickel oxide thin films, Journal of Electroanalytical Chemistry,782,(2016), 36-42, https://doi.org/10.1016/j.jelechem.2016.10.006

[64] V. Srinivasan, J.W. Weidner, An Electrochemical Route for Making Porous Nickel Oxide Electrochemical Capacitors, J. Electrochem. Soc. 144 (1997) L210-L213, https://iopscience.iop.org/article/10.1149/1.1837859/meta

[65] K.R. Prasad, N. Miura, electrochemically deposited nanowhiskers of nickel oxide as a high-power pseudocapacitive electrode, Appl. Phys. Lett. 85 (2004) 4199-4201, https://doi.org/10.1063/1.1814816

[66] D.T. Dam, X. Wang, J.-M. Lee, Mesoporous ITO/NiO with a core/shell structure for supercapacitors, Nano Energy 2 (2013) 1303-1313, https://doi.org/10.1016/j.nanoen.2013.06.011

[67] A. V. Moholkar, S. M. Pawar, K.Y. Rajpure, P.S. Patil ,C.H. Bhosale, Properties of highly oriented spray-deposited fluorine-doped tin oxide thin

films on glass substrates of different thickness, J. Phys. chem. Solids, 68 (2007) 1981, https://doi.org/10.1016/j.jpcs.2007.06.024

[68] Duc Tai Dam, Xin Wang, Jong-Min Lee, Mesoporous ITO/NiO with a core/shell Structure for supercapacitors, Nano Energy, 2(6), (2013),1303-1313, DOI:10.1016/j.nanoen.2013.06.011

[69] Bhim Prasad Kafle, Chemical Analysis and Material Characterization by Spectrophotometry, Paperback ISBN: 9780128148662.

[70] Abhijit Yadav , U.Chavan, Electrochemical Supercapacitive Performance of Spray Deposited NiO Electrodes, Journal of electronic materials, https://doi.org/10.1007/s11664-018-6243-4.

[71] R.S. Kate, S.A. Khalate, R.J. Deokate, Electrochemical properties of spray deposited nickel oxide (NiO) thin films for energy storage systems, J. Anal. Appl. Pyrolysis, (2017)S0165-2370(17)30130-4, http://dx.doi.org/doi:10.1016/j.jaap.2017.03.014.

[72] Gopalakrishnan Srikesh ,Arputharaj Samson Nesaraj, Facile preparation and characterization of novel manganese doped nickel oxide based nanostructured electrode materials for application in electrochemical supercapacitors, (2020),835-847, doi.org/10.1080/21870764.2020.1793477

[73] Xu Han, Bingqing Wang, Can Yang, Ge Meng, Ruifeng Zhao, Qiannan Hu, Okky Triana, Muzaffar Iqbal,Yaping Li, Aijuan Han, Junfeng Liu, Inductive Effect in Mn-Doped NiO Nanosheet Arrays for Enhanced Capacitive and Highly Stable Hybrid Supercapacitor, ACS Appl. Energy Mater. (2019), 2, 3, 2072-2079, https://doi.org/10.1021/acsaem.8b02129

[74] Guohui Yuan, Yunfu Liu, Min Yue, Hongju Li, Encheng Liu, Youyuan Huang, Dongliang Kong, Cu-doped NiO for aqueous asymmetric electrochemical capacitors, 40,7,(2014), 9101-9105, DOI:10.1016/j.ceramint.2014.01.124

[75] Jinping Chen, Xianyun Peng , Lida Song , Lihan Zhang , Xijun Liu , Jun Luo,Facile synthesis of Al-doped NiO nanosheet arrays for high-performance supercapacitors, R Soc Open Sci,5(11),(2018),180842. doi: 10.1098/rsos.180842.

[76] R.S.Kate,R.J.Deokate, Effect of cobalt doping on electrochemical properties of sprayed nickel oxide thin films,3,(2020),860-839, https://doi.org/10.1016/j.mset.2020.06.008

[77] Dewei Liang, Shouliang Wu, Jun Liu, Zhenfei Tiana , Changhao Liang, Co-doped Ni hydroxide and oxide nanosheet networks: laser-assisted synthesis, effective doping, and ultrahigh pseudocapacitor performance, J. Mater. Chem. A, 2016,4, 10609-10617, https://doi.org/10.1039/C6TA03408J

[78] Anjali, P.; Vani, R.; Sonia, T. S.; Nair, A. Sreekumaran; Ramakrishna, Seeram; Ranjusha, R.; Subramanian, K. R. V.; Sivakumar, N.; Mohan, C. Gopi; Nair, Shantikumar V.; Balakrishnan,Avinash, Cerium Doped NiO Nanoparticles: A Novel Electrode Material for High Performance Pseudocapacitor Applications, American Scientific Publishers,6,1,(2014), 94-101(8),DOI: https://doi.org/10.1166/sam.2014.1684

[79] Swati R. Gawali, Deepak P. Dubal, Virendrakumar G. Deonikar, Santosh S. Patil, Seema D. Patil,Pedro Gomez-Romero, Deepak R. Patil, Jayashree Pant,Asymmetric Supercapacitor Based on Nanostructured Ce-doped NiO (Ce:NiO) as Positive and Reduced Graphene Oxide (rGO) as Negative Electrode, Materials Science inc. Nanomaterials & Polymers, 1,13, (2016),3471-3478, DOI: 10.1002/slct.201600566

[80] Lopamudra Halder, Anirban Maitra, Amit Kumar Das, Ranadip Bera, Sumanta Kumar Karan,Sarbaranjan Paria, Aswini Bera, Suman Kumar Si, Bhanu Bhusan Khatua, Fabrication of an Advanced Asymmetric Supercapacitor Based on Three-Dimensional Copper–Nickel–Cerium–Cobalt Quaternary Oxide and GNP for Energy Storage Application, ACS Appl. Electron. Mater. 2019, 1, 189–197, https://doi.org/10.1021/acsaelm.8b00038

[81] Qiu Yang,Zhiyi Lu,Junfeng Liu,Xiaodong Lei,Zheng Chang,Liang Luo,Xiaoming Sun, Progress in Natural Science: Materials International,23, 4, (2013),351- 366, https://doi.org/10.1016/j.pnsc.2013.06.015

Section 6

Redox Type Supercapacitors

Chapter 9

Pseudocapacitors

Thibeorchews Prasankumar, Jemini Jose, Sujin Jose and Sreeja P. Balakrishnan

Abstract

World energy consumption has grown at a rate of knots. Economic growth, increasing prosperity and urbanization, the rise in per capita consumption, and the spread of energy access are the factors likely to considerably increase the total energy demand. In order to meet both the environmental and economic challenges, society realizes the necessity for harvesting the renewable resources, their storage, and recovery. To achieve accelerating clean energy innovation, cost reduction, and deployment of many clean energy technologies, it is important to formulating policies and their implementation, programmes for the development of new and renewable energy apart from coordinating and intensifying R&D in the sector. At present, aggravating energy and environmental issues, such as fossil fuel depletion, pollution problems, and global warming are ringing alarm bells to humans. Thus, there is an urgent need for enhanced energy security along with reducing greenhouse gas emissions. In this direction, renewable energy is one of the environmentally friendly sources of energy and effectiveness of growing economy of the whole world in general. The development of environmentally friendly materials is one of the key issues today.

Keywords: Energy Density, Cycling Stability, Faradaic, Conductivity

1. Introduction

Energy storage devices have received great interest in many emerging modern electronics such as electronic papers, collapsible displays, and other personal multimedia devices. They require highly efficient active materials with good electrochemical properties, high mechanical integrity upon bending, or folding and lightweight property. Renewable energies will not have an anticipated impact unless we find an efficient way to store and use the electricity produced by them [1]. There are various electrochemical energy storage/conversion systems including Capacitors, Supercapacitors (EDLCs, Pseudo-capacitors, Hybrid capacitors), Batteries (Li-ion batteries, Na-ion batteries, Mg-ion batteries), and Fuel cells [2]. Energy storage systems are much required to solve the issues of climate change and storing energy from renewable resources like solar, wind, and biomass. Therefore, energy storage devices have become crucial task for the research community to develop a highly efficient, inexpensive, and eco-friendly energy storage system. Batteries can explore high energy density and deliver low power density that results in the bursts of power requires like the automobiles industry [3]. Supercapacitors, on the other hand, complement batteries and are emerging as highly relevant devices for myriad electronic devices and systems on account of their less charging times (high power density) and exhibit long cycling stability. Hence, they applicable in numerous applications like renewable energy storage, grid leveling, and power conditioning [4, 5].

 IntechOpen

Nanomaterials are having a very important role in the energy storage systems due to their high aspect ratio followed by the enhanced surface area. Energy conversion and storage involve physical and chemical interaction to the surface of the materials. Hence, the surface energy, surface chemistry, and the specific surface area are the key points for these systems [6, 7]. The smaller dimensions of nanomaterials may also offer more favorable mass, heat, and charge transfer, as well as accommodate dimensional changes associated with some chemical reactions and phase transitions [8, 9].

Nanostructured materials deliver extensive surface to volume ratios, commendable transport properties, adjustable physical properties, and confinement effects resulting from the nanoscale dimensions, and have been enormously analyzed for the energy storage applications [10–13]. Metal oxides have always been salient materials that are extensively applicable in different fields, like catalysis, energy storage and conversion, biomedicine, and sensors [14, 15]. Nanostructured metal oxides have attracted tremendous interest in recent years because of their unique physical and chemical properties while the morphological and structural size became down to the nanoscale level [16]. There are lot of metal oxide nanostructures, ranging from nanoparticles, nanowires, nanotubes, and nanoporous structures.

Composite electrodes consist of carbon-based electrode with the combination of metal oxide or conducting polymer in a single electrode. It holds the charge storage mechanisms due to its extensive physical and chemical properties [17, 18]. Through Faradaic reaction, pseudocapacitive material increases the capacitance in the composite electrode [7, 19]. Two types of composite materials are available viz., binary and ternary. Binary composites contain two different electrode materials, whereas in the case of ternary, it comprises three different electrode materials to form a single electrode. In this chapter, the details about pseudocapacitors, the materials involved, and their energy storage performance have been briefly discussed.

2. Supercapacitor

A supercapacitor is a device that stores electrical energy in the electrical double layer that forms at the interface between an electrolyte solution and an electronic conductor [20]. Supercapacitors are one of the energy storage devices that have excessive power density, long cyclic stability, and can be charged and discharged rapidly. They can store more orders of energy than ordinary capacitors. They are friendly to the environment, much safer and can withstand in high temperatures. They have received considerable attention from both academic and industrial arenas because they meet the needs for a wide range of energy storage applications requiring short loading cycles with safety and stability such as power backup systems, portable devices, and electric vehicles. It can be made use of alone or with the combination of fuel cells and batteries. The major advantages of supercapacitors are in hybrid electric vehicles, wearable and flexible electronic gadgets, backup power systems, voltage stabilizer, and power sources for laptops.

Supercapacitors have the same principles as ordinary capacitors (**Figure 1**). However, they must have the electrodes with higher surface areas 'A' and much thinner dielectrics that decrease the distance 'D' between the electrodes [21]. A supercapacitor consists of two electrodes, an electrolyte, and a separator that electrically isolates the two electrodes. These are the essential components that decide the enhanced performance and should be taken into consideration when designing a supercapacitor device. The most important component in a supercapacitor is the electrode material. In general, the supercapacitor electrodes are fabricated from nanoscale materials that have a high surface area and high porosity. The charges are

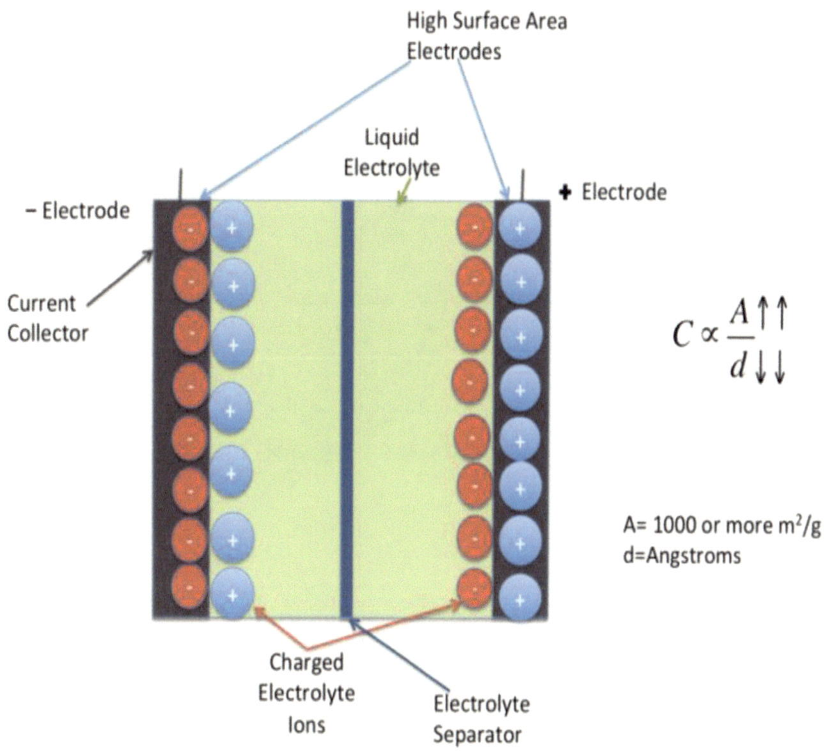

Figure 1.
Schematic of a supercapacitor.

stored and separated at the interface between the conductive solid particles (such as carbon particles or metal oxide particles) and the electrolyte [22]. **Figure 1** explains the schematic of the supercapacitor device [23].

For a two-electrode supercapacitor cell, the two working electrodes are placed across a separator, and the voltage variation in between the electrodes is observed and managed. Each electrode/electrolyte interface represents a capacitor and a resistance. Therefore, the total specific capacitance (C), which is the capacitance per unit mass for the cell, is calculated from the following equation in which C1 and C2 are the specific capacitances of the two electrodes, respectively [24].

$$\frac{1}{C} = \frac{1}{C1} + \frac{1}{C2} \tag{1}$$

The supercapacitors consist of anode, cathode, separator, and electrolyte. The *anode* denotes the negative terminal of a supercapacitor cell associated with the oxidative chemical reactions that expose the electrons into the external circuit. The *cathode* indicates the positive electrode of a cell associated with reductive chemical reactions that observe the electrons from the external circuit. An *electrolyte* is a material that provides pure ionic conductivity between the positive and negative electrodes of a cell. A *separator* is a physical barrier between the positive and negative electrodes that is used to prevent the electrical shorting during the electrochemical reactions. The separator can be a gelled electrolyte or a microporous plastic film, or other porous inert material filled with electrolyte. Separators must be permeable to the ions and inert in the battery environment. The schematic of the pseudocapacitors is explained in **Figure 2** [25].

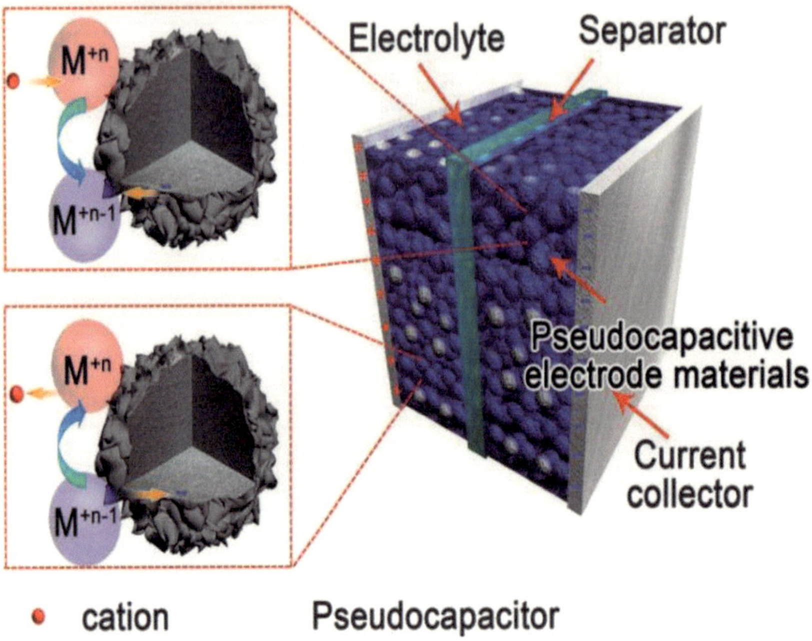

Figure 2.
Schematic of pseudocapacitors.

The Specific capacitance of the supercapacitor cell can be calculated from the charge–discharge measurements using the formula:

$$C = (I.\mathrm{t}) / (\Delta V.\mathrm{m}) \tag{2}$$

where C is the specific capacitance of the electrode material, I is the constant discharge current, t is the discharge time, m is the mass loading of the active material and ΔV is the potential window during discharge. *Energy density* is the amount of energy stored by a supercapacitor.

$$E_s = \frac{1}{2} C \Delta V^2 \tag{3}$$

where 'E_s' is the energy density, 'C' is the specific capacitance of the material and 'ΔV' is the potential window (Volt).

Power density is the amount of power that a supercapacitor can deliver when connected to an external load.

$$P_s = \frac{\mathrm{E_s}}{\mathrm{t}} \tag{4}$$

where 'P_s' is the power density, 'E_s' is the energy density and 't' is the discharge time.

3. Energy storage mechanism in supercapacitors

The performance of the supercapacitor is subjected to the energy storage, the insertion and exertion of ions between the surface of the electrodes, and the

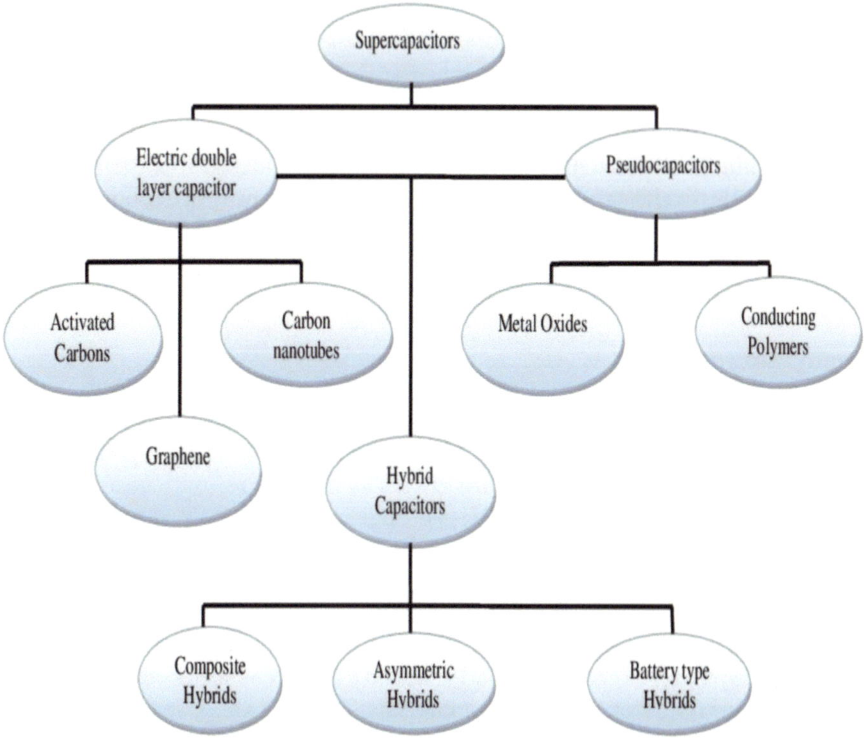

Figure 3.
Taxonomy of supercapacitors.

electrolyte. Based on the energy storage mechanism as well as the active materials used, supercapacitors are classified into three categories: electrochemical double-layer capacitors, pseudocapacitors, and hybrid supercapacitors as shown in **Figure 3**.

Pseudocapacitors store energy through faradaic reaction. They store charge electrostatically in which the transfer of charge between electrode and electrolyte [25]. When a voltage is subjected to a pseudocapacitor, both reduction, and oxidation take place on the electrode material. It involves the passage of charge across the double layer, resulting in faradic current passing through the supercapacitor electrode material. The faradic process employed in pseudocapacitors enhances the electrochemical reactions that result to achieve greater specific capacitance and energy densities compared to EDLCs.

4. Material for pseudocapacitors

4.1 Transition metal oxides (TMO)

Transition-metal oxides, a type of pseudocapacitive material, exhibit fast and reversible redox reactions at the surface of the electrode materials. It exhibits high specific capacitance and low resistance that makes an easy way to construct supercapacitors with high energy and power.

The commonly used metal oxides are RuO_2, MnO_2, Co_2O_3, NiO, SnO_2, Fe_3O_4, IrO_2, V_2O_5, and MoO as supercapacitor electrodes. Among the various metal oxide electrode materials for supercapacitor applications, RuO_2 is considered as the most promising electrode material because its high specific capacitance (720 F g^{-1}) and high electrical conductivity [26]. However, the high cost, as well as the non-availability of Ru on the earth has limited its application in practical usage. By considering these issues, the identification of the low cost pseudocapacitance materials is essential and where the research community is looking into. TMOs exhibit various oxidation states at different potentials and possess crystalline structures that allow high conductivity which enables charges to propagate within their lattice. As faradaic materials, in the same phases, TMOs exhibit two or more oxidation states. When charging and discharging take place, TMOS are able to change their oxidation states, and protons can insert into and extract from the oxide lattice during reduction and oxidation surface reactions.

4.2 Conducting polymers

Conducting polymers (CPs) are the second group of candidate materials for redox pseudocapacitors due to their fast and reversible oxidation/reduction processes, good electrical conductivity, and relatively low cost. The most commonly used conducting polymers include Polyaniline (PANI), Polypyrrole (PPy), Poly (3,4-ethyelene dioxythiophene) (PEDOT), Polythiophene (PTh), and Poly (p-phenylene vinylene) (PPV). They are typically synthesized either through chemical or electrochemical oxidation of the monomer and rendered conductive through a conjugated bond system along the polymer backbone [27, 28]. Conducting polymers exposed the enhanced conductivity, capacitance, and reduced equivalent series resistance when compared with carbon-based electrode materials. The ions are observed/drifted from the electrolyte to the polymers and released to the electrolyte while oxidation and reduction take place correspondingly. Because of no phase transition occurs, the conductive polymers exhibit highly reversible reactions that lead to the enhanced cycling stability. They are positively or negatively charged by the redox reactions that induce the enhanced conductivity.

4.3 Composites of conducting polymers with metal oxides/hydroxides

The electrode materials of metal hydroxides/oxides have better energy densities than carbon materials and higher cyclic stability than conductive polymers [29]. Besides, the higher capacitances of metal hydroxides/oxides are due to the electrochemical process which occurs on the surface and bulk of the electrode materials [30]. The composites of metal hydroxides/oxides with conducting polymer perform remarkable cyclic stability and high specific capacitance that reduce the agglomeration and the restacking of metal hydroxide/oxide permitting uniform dispersion of metal hydroxides/oxides and the conducting polymer matrix [31–33]. Moreover, the association between the electrode and current collector increases the electroactive area in metal oxide/hydroxide-conducting polymer composites. The low cost and high theoretical capacitance of MnO_2 (1100 F g^{-1}) make it as a good alternative for expensive metal oxides and its composites with CPs have become one of the centre of attraction of many researchers. Sharma et al., synthesized nucleated MnO_2 nanoparticles over polypyrrole chains by electrochemical co-deposition in polished graphite surfaces. The embedded PPy chains of MnO_2 offer high surface area due to the special arrangement of MnO_2 nanoparticles and PPy in the composite. The nanocomposite showed better Faradaic redox reactions which in turn lead to significant supercapacitor performance with good conductivity and stability. The specific capacitance obtained was 620 F g^{-1} which is higher than its individual components MnO_2 (225 F g^{-1}) and PPy (250 F g^{-1}) [34]. Zhang et al. intercalated PANI-MnO_2 nanocomposite by exchange reaction with PANI and MnO_2 intercalated n-octadecyltrimethylammonium precursors using N-methyl-2-pyrrolidone as the solvent. The nanocomposite exhibited the highest specific capacitance of 330 F g^{-1} with 94% of capacitance retention even after 1000 cycles. The composite exhibited outstanding performance than MnO_2 and PANI [35]. Liu et al., introduced the synthesis of nanocomposite MnO_2/PEDOT by electro co-deposition of EDOT monomer (80 mM) and manganese acetate (10 mM) in a porous alumina template. The coaxial nanowires of the composite exhibited high specific capacitance (210 F g^{-1}) and high current density due to the fast ionic diffusion into the core MnO_2 of the coaxial nanowires [36]. Gan et al., synthesized MnO_2/PPy composite by in situ chemical synthesis method which improves the porosity and enhances specific surface area with fast intercalation/deintercalation of the electrolyte. The fabricated supercapacitor device showed the specific capacitance of 312 F g^{-1} at 10 mV s^{-1} with 93.2% cyclic stability over 1000 charge/discharge cycles [37]. Debelo et al., prepared single-phase electrodeposition on stainless steel using the aqueous solution of manganese nitrate and pyrrole. The $Mn(OH)_2$/PPy porous nanosheets exhibited the specific capacitance of 430 and 220 F g^{-1} for anodic and cathodic films [38]. Some of the MnO_2 and PPy composites are shown in **Figure 4** [37, 38].

Along with the CPs with MnO_2, nickel hydroxide ($Ni(OH)_2$) and nickel oxide (NiO) form composites with CPs as supercapacitor electrode material owing to the easy synthetic procedure, low cost, and redox reactions, and high theoretical specific capacitance (2584 F g^{-1} for NiO and 2081 F g^{-1} for $Ni(OH)_2$) [39]. Sun et al., used binder free approach for the synthesis of NiO-PANI on nickel foam for the fabrication of supercapacitor. The composite showed 2565 F g^{-1} specific capacitance at 1 A g^{-1} and showed excellent cyclic stability with 100% retention of specific capacitance after 5000 cycles [40]. Yang et al., encapsulated porous

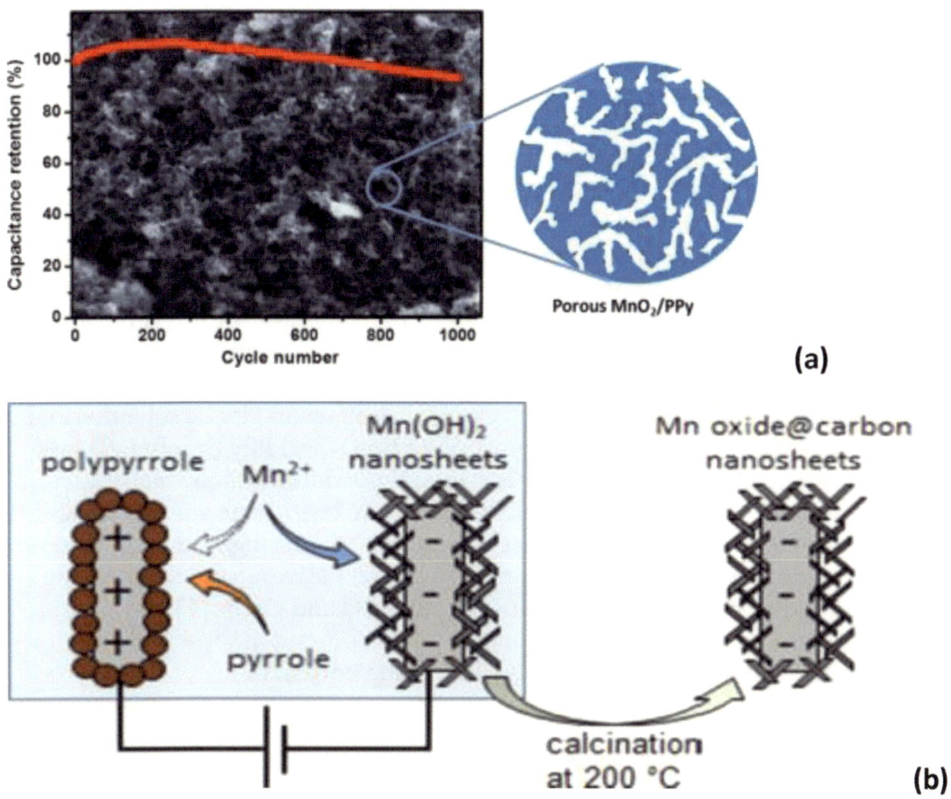

Figure 4.
MnO2 with PPy composites for supercapacitor (a and b).

$NiO/Ni(OH)_2$ with PEDOT on Cu-Ni alloy wires by facile and low cost electrochemical route. The protective layer of PEDOT on the surface of PNS enhances the electronic conductivity and stability. This hybrid electrode achieved a high specific capacitance of 404.1 mF cm^{-2} at a current density of 4 mA cm^{-2} with 82.2% capacitance retention even after 1000 cycles. The supercapacitor device delivered the power density of 0.33 mW cm^{-2} and the energy density of 0.011 mWh cm^{-2} with an output voltage of 1.5 V [41].

Vanadium oxide (V_2O_5), cobalt monoxide (CoO), and hematite (a-Fe_2O_3) have also been explored as supercapacitor electrode materials with CPs. The layered structure, different oxidation state, and redox reactions at the electrode surface make it as a good energy material [29]. Liu et al., synthesized V_2O_5-PANI composite by electro co-deposition method with a specific capacitance of 443 F g^{-1} at the potential window 1.6 V. The composite film exhibited a power density of 720 W kg^{-1} and an energy density of 69.2 Wh kg^{-1} with excellent stability [42]. The same group electro-codeposited V_2O_5-PPy using an aqueous solution of vanadyl sulphate ($VOSO_4$) and pyrrole. The nanocomposite showed specific capacitance 412 F g^{-1} at 4.5 mA cm^{-2} owing to the synergistic effect of V_2O_5 and PPy.

It exhibited the operating voltage window of 2 V with an energy density of 82 Wh kg^{-1} [32]. Aamir et al., fabricated an electro-codeposition technique for the synthesis of V_2O_5-PANI composite on Ni foam which is a porous and conductive framework for pseudocapacitors. The synthesized composite showed the specific capacitance of 1115 F g^{-1} due to the small charge transfer resistance [43]. α -Fe_2O_3 has also been investigated as the electrode material for supercapacitors because of its abundance, low cost, and good theoretical capacitance. The inclusion of α-Fe_2O_3 on CPs makes α-Fe_2O_3 as a promising material for supercapacitance properties with good cyclic stability. Lu et al., investigated the areal capacity of α-Fe_2O_3@PANI core shell synthesized by simple electrodeposition method. The electrochemical performance of the nanowires arrays offers enhancement of structure stability, fast movement, and transfer of ions, and large surface area. The designed material serves as the anode and PANI on carbon cloth as cathode and obtained good volumetric capacitance (2.02 mF cm^{-3}). The fabricated asymmetric supercapacitor also experienced a high power density (120.51 mW cm^{-3}) and energy density (0.35 mW cm^{-3}) with excellent cyclic stability [44].

In order to address the capacitance and stability issues, CoO was also incorporated with CPs for pseudocapacitor. Liu et al., developed CoO@PPy nanowire on nickel foam to enhance the pseudocapacitor due to the synergistic effect of CoO and PPy. The well aligned asymmetric supercapacitor showed a high specific capacitance (2223 F g^{-1}) with capacitance retention of 99.8% after 2000 cycles [45]. Yang et al., fabricated a hybrid electrode with Co_3O_4 nanosheet on PPy by solvothermal route and electrodeposition. The combination of Co_3O_4 and PPy benefits the fast transfer of ions at the electrode-electrolyte interface exhibited high areal capacitance (2.11 F cm^{-2}) at the current density of 2 mA cm^{-2} with superior cyclic stability (after 5000 cycles) of 85.5% capacitance retention [46]. The high conductivity and electrochemical activity of nickel cobaltite ($NiCo_2O_4$) make outstanding specific capacitance (1400 F g^{-1}) in comparison with pure NiO and Co_3O_4 [47, 48].

4.4 Composites of metal sulfides with conducting polymers

Metal sulfides are cheap and abundant in nature undergo different valence states and redox reactions due to the presence of metal ions [49]. The metal sulfides of Mo, Cu, Ni, and Co with CPs have become the centre of attraction due to the high specific capacitance and outstanding reversible redox reactions [50]. Peng et al., developed CuS with PPy by solvothermal route in the absence of surfactant and exhibited the specific capacitance of 427 F g^{-1} with outstanding cyclic stability due to the intertwined structural properties of CuS@PPy [51]. Wei et al., synthesized flexible electrode with PPy/NiS/bacterial cellulose and obtained excellent supercapacitor properties with 713 F g^{-1} specific capacitance at 0.8 mA cm^{-2} [52]. MoS_2 is considered as one of the excellent metal sulfides with high electrochemical capacitive properties and variable oxidation state of Mo (+2 to +6) which is a promising material for pseudocapacitance behavior. Ren et al., used in situ oxidative polymerization method for the controllable synthesis of MoS_2/PANI hybrid material where PANI nanowires are aligned on the external and internal surface of MoS_2.The porosity and excellent conductivity of the hybrid material provide fast ion diffusion between the electrode and the electrolyte. The capacity to withstand the volume change during cycling and the large surface area, make it as a smart material for the excellent specific capacitance of 552 F g^{-1} at 0.5 A g^{-1} and rate capability of 82% from 0.5 to 30 A g^{-1} [53]. Yang et al., synthesized carbon shell coated PANI grown on 1 T monolayers of MoS_2 (MoS_2/PANI@C). The composite exhibited outstanding specific capacitance of 678 F g^{-1} in 1 mV s^{-1} with excellent capacitance retention of 80% even after 10,000 cycles due to synergistic effect of PANI,1 T MoS_2, as well as the thin carbon shell [54].

5. Conclusion

The development of novel materials for energy applications has been identified as one of the priority areas of all the other countries. In order to meet both the environmental and economic challenges, the globe realizes the necessity for harvesting renewable resources, their storage, and recovery. We believe that this generation holds the key to transforming the future of sufficient energy for usage, which could be possible by making improvements in energy storage devices. By considering these issues, the energy storage mechanism of supercapacitors and the details about the advanced materials of pseudocapacitors with some recent literature works are briefly discussed in this chapter for the researchers to develop novel pseudocapacitors.

Acknowledgements

The authors would like to acknowledge Department of Science and Technology, Government of India for the financial support through DST-MES project (DST/TMD/MES/2 K17/94).

Author details

Thibeorchews Prasankumar[1]*, Jemini Jose[2], Sujin Jose[1] and Sreeja P. Balakrishnan[2]

1 School of Physics, Madurai Kamaraj University, Madurai, Tamil Nadu, India

2 Department of Chemistry, CHRIST (Deemed to be University), Bengaluru, Karnataka, India

*Address all correspondence to: prasankumar04@gmail.com

References

[1] Arumugam Manthiram, Materials Challenges and Opportunities of Lithium Ion Batteries, Physical Chemistry Letters. 2 (2011) 176-184.

[2] John B. Goodenough, Kyu-Sung Park, The Li-Ion Rechargeable Battery: A Perspective, Journal of American Chemical Society. 135 (2013) 1167-1176.

[3] HaishengChen, Yulong Ding, Yongliang Li, Xinjing Zhang, Chunqing Tan, Air fuelled zero emission road transportation: A comparative study, Applied Energy. 88 (2011) 337-342.

[4] Yang-Wu Shen; De-Ping Ke; Yuan-Zhang Sun; Daniel S. Kirschen; Wei Qiao; Xiang-Tian Deng, Advanced Auxiliary Control of an Energy Storage Device for Transient Voltage Support of a Doubly Fed Induction Generator, IEEE Transactions on Sustainable Energy. 7 (2015) 63-76.

[5] Bruce Dunn, Haresh Kamath, Jean-Marie Tarascon, Electrical Energy Storage for the Grid: A Battery of Choices, Science. 334 (n.d.) 928-935.

[6] Vaibhav Lokhande, Supercapacitive composite metal oxide electrodes formed with Carbon, Metal Oxides and Conducting Polymers, Journal of Alloys and Compounds. 682 (2016) 381-403.

[7] JeminiJose, Sujin Jose, S.Abinaya, Sadasivan Shaji, P.B.Sreeja, Benzoyl hydrazine-anchored graphene oxide as supercapacitor electrodes, Material Chemistry and Physics. 256 (2020) 123666.

[8] Dawei Liu, Guozhong Cao, Engineering nanostructured electrodes and fabrication of film electrodes for efficient lithium ion intercalation, Energy Environ. Sci. 3 (2010) 1218-1237.

[9] Bruno Scrosati, Jürgen Garche, Lithium batteries: Status, prospects and future, Journal of Power Sources. 195 (2010) 2419-2430.

[10] Navneet Sharma, Himanshu Ojha, Ambika Bharadwaj, Dharam Pal Pathak, Rakesh Kumar Sharma, Preparation and catalytic applications of nanomaterials: a review, RSC Advances. 5 (2015) 53381-53403.

[11] Lina Cheng, Zhi-Gang Chen, Lei Yang, Guang Han, Hong-Yi Xu, G. Jeffrey Snyder, Gao-Qing Lu, Jin Zou, T-Shaped Bi_2Te_3–Te Heteronanojunctions: Epitaxial Growth, Structural Modeling, and Thermoelectric Properties, Journal of Physical Chemistry C. 117 (2013) 12458-12464.

[12] Chunhai Jiang, Eij Hosono, Haoshen Zhou, Nanomaterials for lithium ion batteries, Nano today. 1 (2006) 28-33.

[13] Sanjit Saha, Pranab Samanta, Naresh Chandra Murmu, Tapas Kuila, A review on the heterostructure nanomaterials for supercapacitor application, Journal of Energy Storage. 17 (2018) 181-202.

[14] Thibeorchews Prasankumar, Smagul Karazhanov, Sujin P. Jose, Three-dimensional architecture of tin dioxide doped polypyrrole/reduced graphene oxide as potential electrode for flexible supercapacitors, Material Letters. 221 (2018) 179-182.

[15] Sibi Abraham, T .Prasankumar, Vinoth Kumar, Smagul Zh Karazhanov, Sujin Jose, Novel lead dioxide intercalated polypyrrole/graphene oxide ternary composite for high throughput supercapacitors, Material Letters. 273 (2020) 127943.

[16] Jemini Jose, Sujin P.Jose, T.Prasankumar, Sadasivan Shaji, Saju Pillai, Sreeja P.B, Emerging ternary nanocomposite of rGO draped

palladium oxide/polypyrrole for high performance supercapacitors, Journal of Alloys and Compounds. 855 (2021) 157481.

[17] Sujin P. Jose, Chandra Sekhar Tiwary, Suppanat Kosolwattana,a Prasanth Raghavan, Leonardo D. Machado, Chandkiram Gautam, T. Prasankumar, Jarin Joyner, Sehmus Ozden, Douglas S. Galvaoc, P. M. Ajayan, Enhanced supercapacitor performance of a 3D architecture tailored using atomically thin rGO–MoS2 2D sheets, RSC Adv. 6 (2016) 93384-93393.

[18] Ibrahim Khan, Khalid Saeed, IdreesKhan, Nanoparticles: Properties, applications and toxicities, Arabian Journal of Chemistry. 12 (2019) 908-931.

[19] Aimee M. Bryan, Luciano M. Santino, Yang Lu, Shinjita Acharya, Julio M. D'Arcy*‡§, Conducting Polymers for Pseudocapacitive Energy Storage, Chemistry of Materials. 28 (2016) 5989-5998.

[20] John R. Miller and Andrew Burke, Electrochemical Capacitors: Challenges and Opportunities for Real-World Applications, The Electrochemical Society Interface. 17 (2008) 1.

[21] B.E. Conway, Plenum Press, Kluwer Academic Publishers, Newyork, 1999.

[22] Jintao Zhang , X. S. Zhao, On the Configuration of Supercapacitors for Maximizing Electrochemical Performance, CHEMSUSCHEM. 5 (2012) 818-841.

[23] Martin Winter and Ralph J. Brodd, What Are Batteries, Fuel Cells, and Supercapacitors?, Chem.Rev. 104 (2004) 4245-4270.

[24] MichioInagakia Hidetak, Konno, Osamu Tanaike, Carbon materials for electrochemical capacitors, Journal of Power Sources. 195 (2010) 7880-7903.

[25] Cheng Zhong, Yida Deng, Wenbin Hu, Jinli Qiao, Lei Zhang, Jiujun Zhang, A review of electrolyte materials and compositions for electrochemical supercapacitors, Chemical Society Reviews. 44 (2015) 7484-7539.

[26] Chi-Chang Hu, Kuo-Hsin Chang, Ming-Champ Lin, and Yung-Tai Wu, Design and Tailoring of the Nanotubular Arrayed Architecture of Hydrous RuO_2 for Next Generation Supercapacitors, Nano Letters. 6 (2006) 2690-2695.

[27] Jian Jiang, Yuanyuan Li, Jinping Liu Xintang Huang Changzhou Yuan Xiong Wen (David) Lou, Recent Advances in Metal Oxide-based Electrode Architecture Design for Electrochemical Energy Storage, Advanced Materials. 24 (2012) 5166-5180.

[28] *D. Krishna* Bhat & *M. selva* Kumar, N and p doped poly(3,4-ethylenedioxythiophene) electrode materials for symmetric redox supercapacitors, Journal of Material Science. 42 (n.d.) 8152-8162.

[29] C.D.Lokhande, D.P.Dubal, Oh-Shim Joo, Metal oxide thin film based supercapacitors, Current Applied Physics. 11 (2011) 255-270.

[30] Xingyou Lang, Akihiko Hirata, Takeshi Fujita & Mingwei Chen, Nanoporous metal/oxide hybrid electrodes for electrochemical supercapacitors, Nature Nanotechnology. 6 (2011) 232-236.

[31] Dan-Dan Zhao, Shu-Juan Bao, Wen-Jia Zhou, Hu-Lin Li, Preparation of hexagonal nanoporous nickel hydroxide film and its application for electrochemical capacitor, Electrochemistry Communications. 9 (2007) 869-874.

[32] Ming-Hua Bai, Li-Jun Bian, Yu Song, and Xiao-Xia Liu, Electrochemical Codeposition of Vanadium Oxide and

Polypyrrole for High-Performance Supercapacitor with High Working Voltage, ACS Applied Materials and Interfaces. 6 (2014) 12656-12664.

[33] Shuang Li Dr. Dongqing Wu Chong Cheng Jinzuan Wang Dr. Fan Zhang Dr. Yuezeng Su Prof. Xinliang Feng, Polyaniline-Coupled Multifunctional 2D Metal Oxide/Hydroxide Graphene Nanohybrids, Angewandte Chemie. 52 (2013) 12105-12109.

[34] Raj Kishore Sharma, Alok Rastogi, Sheshubabu Desu, Manganese oxide embedded polypyrrole nanocomposites for electrochemical supercapacitor, Electrochimica Acta. 53 (n.d.) 7690-7695.

[35] Xiong Zhang, Liyan Ji, Shichao Zhang, Wensheng Yang, Synthesis of a novel polyaniline-intercalated layered manganese oxide nanocomposite as electrode material for electrochemical capacitor, Journal of Power Sources. 173 (2007) 1017-1023.

[36] Ran Liu and Sang Bok Lee, MnO2/ Poly(3,4-ethylenedioxythiophene) Coaxial Nanowires by One-Step Coelectrodeposition for Electrochemical Energy Storage, Journal of American Chemical Society. 130 (2008) 2942-2943.

[37] John Kevin Gan, Yee Seng Lim,Nay Ming Huang, Hong Ngee Lim, Effect of pH on morphology and supercapacitive properties of manganese oxide/ polypyrrole nanocomposite, 357 (2015) 479-486.

[38] Tamene Tamiru Debelo, Masaki Ujihara, Effect of simultaneous electrochemical deposition of manganese hydroxide and polypyrrole on structure and capacitive behavior, Journal of Electroanalytical Chemistry. 859 (2020) 113825.

[39] Anne-Lise Brisse,Philippe Stevens, Gwenaëlle Toussaint, Olivier Crosnier, Thierry Brousse, $Ni(OH)_2$ and NiO Based Composites: Battery Type Electrode Materials for Hybrid Supercapacitor Devices, Materials. 11 (2018) 1178.

[40] Bangning Sun, Xinping He, Xijin Leng, Yang Jiang, Yudong Zhao, Hui Suo and Chun Zhao, Flower-like polyaniline–NiO structures: a high specific capacity supercapacitor electrode material with remarkable cycling stability, RSC Adv. 6 (2016) 43959-43963.

[41] Huiling Yang, Henghui Xu, Ming Li, Lei Zhang, Yunhui Huang, and Xianluo Hu, Assembly of NiO/$Ni(OH)_2$/PEDOT Nanocomposites on Contra Wires for Fiber-Shaped Flexible Asymmetric Supercapacitors, ACS Applied Materials and Interfaces. 8 (2016) 1774-1779.

[42] Ming-Hua Bai,Tian-Yu Liu,Feng Luan,Yat Li , Xiao-Xia Liu, Electrodeposition of vanadium oxide–polyaniline composite nanowire electrodes for high energy density supercapacitors, Journal of Material Chemistry A. 2 (2014) 10882-10888.

[43] Asma Aamir, Adil Ahmad, Said Karim Shah, Noorul Ain, Mazhar Mehmood, Yaqoob Khan, Zia ur Rehman, Electro-codeposition of V_2O_5-polyaniline composite on Ni foam as an electrode for supercapacitor, Journal of Materials Science: Materials in Electronics. 31 (2020) 21035-21045.

[44] Xue-Feng Lu, Xiao-Yan Chen, Wen Zhou, Ye-Xiang Tong, Gao-Ren Li, α-Fe_2O_3@PANI Core-Shell Nanowire Arrays as Negative Electrodes for Asymmetric Supercapacitors, ACS Applied Materials and Interfaces. 7 (2015) 14843-14850.

[45] Cheng Zhou, Yangwei Zhang, Yuanyuan Li, Jinping Liu, Construction of High-Capacitance 3D CoO@ Polypyrrole Nanowire Array Electrode for Aqueous Asymmetric

Supercapacitor, ACS Applied Materials and Interfaces. 13 (2013) 2078-2085.

[46] Xiaojun Yang. Kaibing Xu, Rujia Zou. Junqing Hu, A Hybrid Electrode of Co_3O_4@PPy Core/Shell Nanosheet Arrays for High-Performance Supercapacitors, Nano-Micro Letters. 8 (2016) 143-150.

[47] Zhibin Wu,Yirong Zhu, Xiaobo Ji, $NiCo_2O_4$-based materials for electrochemical supercapacitors, Journal of Material Chemistry A. (2014) 14759-14772.

[48] Nawishta Jabeen, Qiuying Xia, Mei Yang, and Hui Xia, Unique Core–Shell Nanorod Arrays with Polyaniline Deposited into Mesoporous $NiCo_2O_4$ Support for High-Performance Supercapacitor Electrodes, ACS Applied Materials and Interfaces. 8 (2016) 6093-6100.

[49] Chen-Ho Lai , Ming-Yen Lu and Lih-Juann Chen, Metal sulfide nanostructures: synthesis, properties and applications in energy conversion and storage, Journal of Material Chemistry. 22 (2012) 19-30.

[50] Jiaqin Yang, Xiaochuan Duan, Qing Qina, Wenjun Zheng, Solvothermal synthesis of hierarchical flower-like β-NiS with excellent electrochemical performance for supercapacitors, Journal of Material Chemistry A. 27 (2013) 7880-7884.

[51] Hui Peng,Guofu Ma, Kanjun Sun, Jingjing Mu, Hui Wang, Ziqiang Lei, High-performance supercapacitor based on multi-structural CuS@polypyrrole composites prepared by in situ oxidative polymerization, Journal of Material Chemistry A. 2 (2014) 3303-3307.

[52] Shuo Peng, Lingling Fan, Chengzhuo Wei, Haifeng Bao, Hongwei Zhang .Weilin Xu .Jie Xu, Polypyrrole/ nickel sulfide/bacterial cellulose nanofibrous composite membranes for flexible supercapacitor electrodes, Cellulose. 23 (2016) 2639-2651.

[53] Lijun Ren, Gaini Zhang, Zhe Yan, Liping Kang, Hua Xu, Feng Shi, Zhibin Lei, and Zong-Huai Liu, Three-Dimensional Tubular MoS_2/PANI Hybrid Electrode for High Rate Performance Supercapacitor, ACS Applied Materials and Interfaces. 7 (2015) 28294-28302.

[54] Chao Yang, Zhongxin Chen, Imran Shakir, Yuxi Xu, Hongbin Lu, Rational synthesis of carbon shell coated polyaniline/MoS_2 monolayer composites for high-performance supercapacitors, Nano Research. 9 (2016) 951-962.

Section 7

Supercapacitors Designs

Chapter 10

Most Modern Supercapacitor Designs Advanced Electrolyte and Interface

Yachao Zhu and Olivier Fontaine

Abstract

Electrolyte plays a key and significant role in supercapacitors. The interaction of an electrode and a chosen electrolyte has a significant effect on the parameters., i.e., ionic conductivity, stable potential range, and charge transfer coefficient, therefore determining the corresponding performance. The captivating interface between electrode and electrolyte is also pushing the intensive research. In this chapter, we focus on two kinds of electrolytes, including water-in-salt electrolytes and redox-ionic liquid. Water-in-salt electrolyte is drawing continuous attention thanks to the formed hydrophobic layer on the positive electrode and solid electrolyte interphase (SEI) on the negative side, preventing water splitting. On the other side, redox-ionic liquid, taking advantage of the broad and stable working window, on the interface, the redox shuttle passes and targets the suitable electrode bulk, leading to redox reactions to highlight capacitance and energy.

Keywords: electrolyte, water-in-salt, redox-ionic liquid, interface, supercapacitor

1. Introduction

Apart from the electrodes [1–4], the essential electrolyte is also one of the most critical components to building the advanced supercapacitor, which involves electrochemical stable voltage window, capacitance, power density, lifetime, stability, and safety [5–7]. According to the calculation of energy, energy density is proportional to the square of the cell voltage, suggesting that it is more influential in enlarging the voltage window, which is decided mainly by the nature of applied electrolytes [8, 9]. As a critical parameter, to match the applied electrode material, the selected electrolyte has a significant impact on the interface between electrode and electrolyte, ionic conductivity, and charge transfer, therefore determining the performance of supercapacitor [10]. As shown in **Figure 1**, the publications number of electrolyte research largely increases over the past decades, signifying its promising research potential. Usually, the electrolyte is composed of salt and solvent. In the light of the classification tree of electrolytes, generally, the electrolyte includes the liquid electrolyte, solid-state or quasi-solid-state electrolyte, and redox-active electrolyte. Liquid electrolyte contains aqueous, non-aqueous, and water-in-salt electrolytes. Redox-active electrolytes include aqueous, ionic liquid, and some gel polymer electrolytes (**Figure 2**).

There into, the highlighted merits of wide working voltage window, adsorbing solid-electrolyte interphase, or enhanced redox moiety earn considerable attention

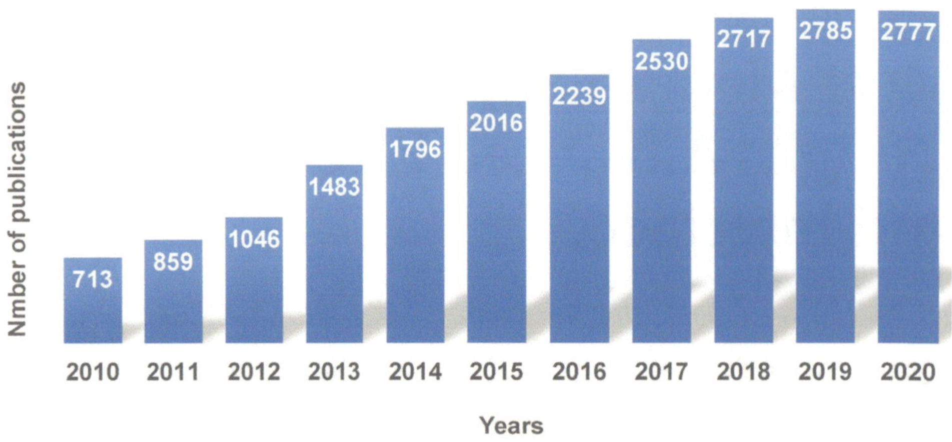

Figure 1.
The statistics of publications of electrodes and electrolytes. Search formulation: ("supercapacitor" or "electrochemical capacitor*" or "ultracapacitor*" or "double-layer capacitor*" or "pseudocapacitor*"); searched from web of science; search time: 15, Feb 2021.*

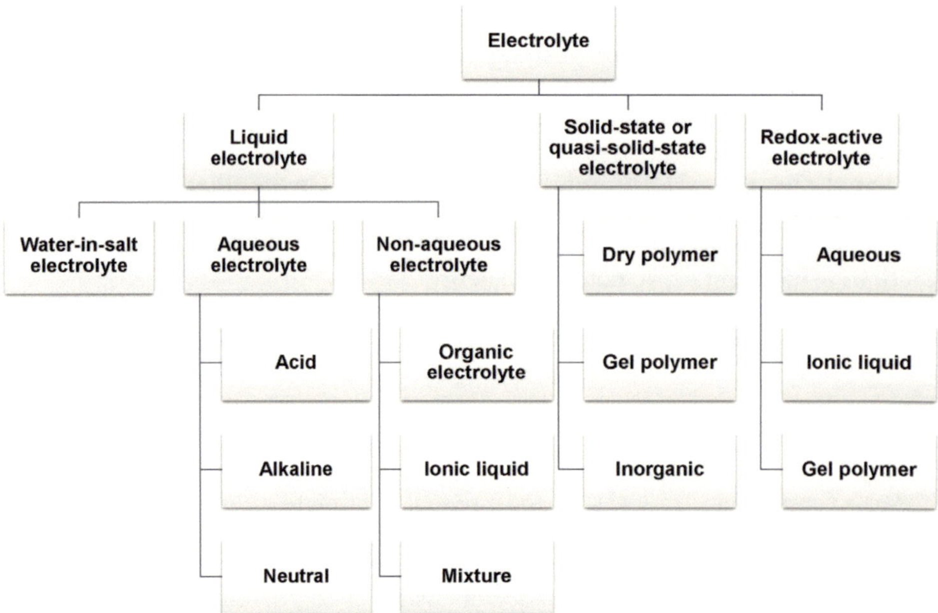

Figure 2.
The classification tree of electrolytes for supercapacitor.

on water-in-salt electrolyte [11] and redox-ionic liquid electrolyte [12]. Despite safe, green, and low-cost, aqueous electrolyte already reaches the ceiling of limited voltage (1.23 V), which impedes the practical applications in high-energy supercapacitor [13–15]. Based on the definition of energy, the working voltage window plays a vital role in boosting the energy. In this regard, except replacing with non-aqueous electrolyte, water-in-salt electrolyte, breaking through the thermodynamic stability limits of water, is prevailing in energy storage [16–20]. As followed main standpoint, for the positive electrode, anions are forced to aggregate at the interface, thus shaping a dense hydrophobic layer to insulate the water from the electrode and further pushing more positive potential. For the negative side, it is agreed that a solid electrolyte interphase (SEI) forms to suppress the hydrogen evolution reaction (HER), thereby bringing downward more negative potential. It is found that the SEI can block the electronic conduction, but the ions can still pass [21–23]. However, the formation mechanism is still controversial, especially considering the period and the way of decomposing TFSI anions to form SEI, together with water decomposition. Besides, to promote energy, another strategy is to stimulate capacitance. Except scheming new intricate electrode materials, a redox-ionic liquid electrolyte [12, 24–27] leads to an explosion of research in practical applications, such as supercapacitor. Initially, the inert ionic liquid has many good merits, including a wide working voltage window, low vapor pressure, non-flammability, and good thermal stability [28–31]. Building the primary ionic liquid with active redox species can result in an enhanced electric double-layer capacitor, on/near the interface, occurring ions adsorption/desorption, and fast reversible redox reactions [32]. Taking advantage of the wide working window and redox moiety contributions can motivate higher energy for supercapacitor. But it is worth noting that the mechanism is still open in the case of a supercapacitor containing redox shuttles on/near the interface, mainly varying different redox components.

2. Water-in-salt electrolyte

Depended on the malleable potential window, the water-in-salt electrolyte has been extensively reported in numerous papers thanks to SEI's great stability. It is significant to figure out the basic fundamental of SEI formation for further practical applications.

2.1 The concentration effect of water-in-salt electrolyte

Water-in-salt electrolyte, once reported, has been widely researched in lithium-ion batteries and supercapacitor. As shown in **Figure 3**, salt-in-water, exhibits a conventional solvation sheath structure, including a primary solvation sheath and loose-bound secondary solvation sheath. The anion and cation are isolated by water. Compared with the salt-in-water part, the essence of water-in-salt is that the high-soluble-capability salt with extra high concentration (more than 20 m) is dissolved in water, and the insufficient water is not able to neutralize the electrostatic field of Li^+, thus resulting in anion-containing-cation solvation sheath, where it occurs the corresponding interactions between cation and anion. Initially, a series of concentrated

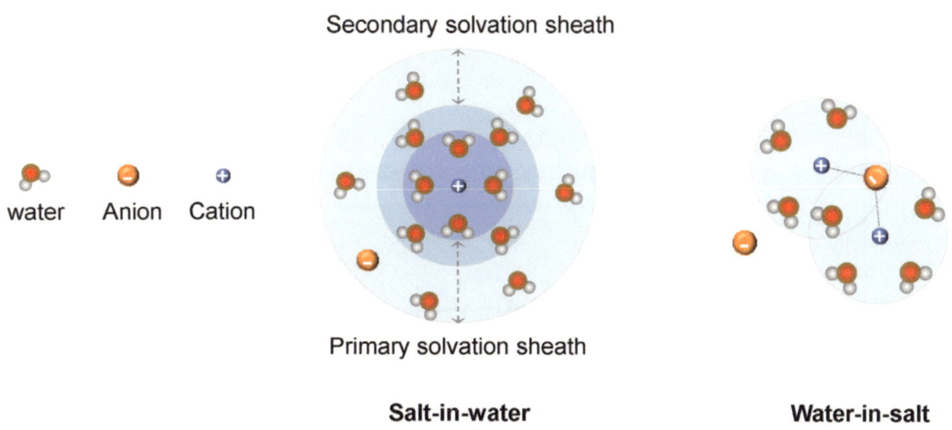

Figure 3.
The difference between salt-in-water and water-in-salt electrolytes.

lithium bis(trifluoromethyl)sulfonimide (LiTTSI) electrolytes were first reported by Suo et al. [33] For various concentrations of LiTSFI, cyclic voltammetry (CV) program was used to evaluate the electrochemical stability window, which is depicted in **Figure 4**. On the positive electrode, increasing the concentrations, it displayincreasing the concentrations show that the oxygen evolution process is gradually suppressed until having an onset potential of 4.9 V at 21 m (**Figure 4B**). On the positive part, for hydrogen evolution, a passivation process is found as increasing the concentration, which visualizes the plateaus, further pushing the onset potential more downward, namely, from 2.63 V to 1.90 V (**Figure 4A**). Extending the potential operating window is remarkable, compelling to figure out the story on the electrode surface.

2.2 The principles of water-in-salt electrolyte

The story on the electrode surface is captivating to find out in water-in-salt electrolytes. Numerous studies have been conducted to shed light on the mechanism underlying the broadening of the operating voltage window, employing a variety of characterization techniques. First, on the positive surface, although it is almost certain that TFSI anions combine to form a dense hydrophobic layer that separates the water from the electrode, the ions remain mobilizable. The debate mainly focuses on the negative side, where the formation mechanism of solid electrolyte interphase (SEI) is still controversial. The SEI existence efficiently obstructs the water approaching the electrode's surface, but lithium ions still pass quickly,

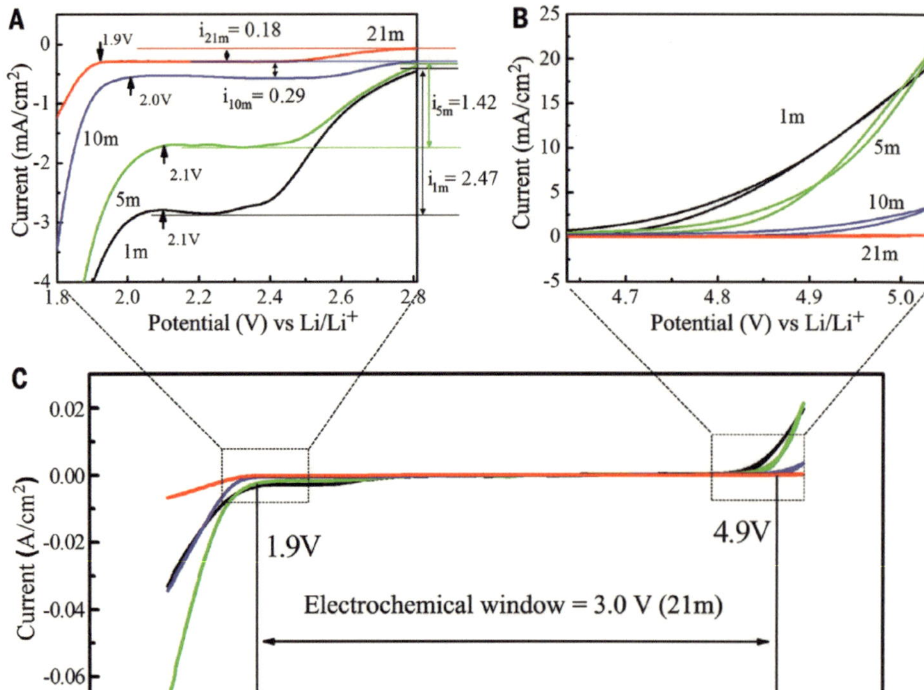

Figure 4.
The electrochemical stability window of LiTFSI-H_2O electrolytes with various molality on nonactive electrodes. (A) and (B) Magnified view of the regions of outlined on negative and positive sides, (C) overall electrochemical stability window. This figure has been adapted from Ref. [33] with permission from Science.

thus suppressing the hydrogen evolution reaction (HER). One of the authentic explanations is from Dubouis et al. [34], as shown in **Figure 5**: initially water is reduced starting from the onset potential, and OH^- is released along with H_2. Then OH^- bonds with Li^+ to have LiOH, which is deposited on the surface. After that, a nucleophilic attack occurs between LiOH or OH^- and TFSI anions (electrophilic sulfur, precisely), generating F^- and some organic compounds, which bond with Li^+ again to form a fluorinated layer, i.e., SEI. Significantly, the authors demonstrated that the decomposition procedure of water and TFSI is proceeding simultaneously. Bouchal et al. [35] disclosed a precipitation/dissolution model, meaning that the SEI is not stable following continuous dissolution due to local oversaturation at the interface. The calculated model disclosed that the dissolution stems from the driving force, which results from adverse flows between water (from the solution to the electrode) and ions (from the electrode to the solution). This model is versatile to interpret the SEI formation and then the dynamic decomposition/regeneration of SEI layer in a high-concentration electrolyte. Furthermore, they demonstrated that over reducing free water process, the TFSI is chemically decomposed, while chemical and electrochemical degradations of TFSI occur over reducing bound water.

Water-in-salt definition sets the stage for extending the operating voltage window in aqueous electrolytes, which means that a series of advanced electrode materials in water can still work with an organic-electrolyte-like voltage arrange, therefore targeting the high energy density but still keeping safe, green, and low-cost properties. However, some challenge is still there. The critical issue is that it is not passivated enough to protect the whole electrode surface from water at the first cycle, as shown in **Figure 6** [36]. It is found that the passivation is a gradual process,

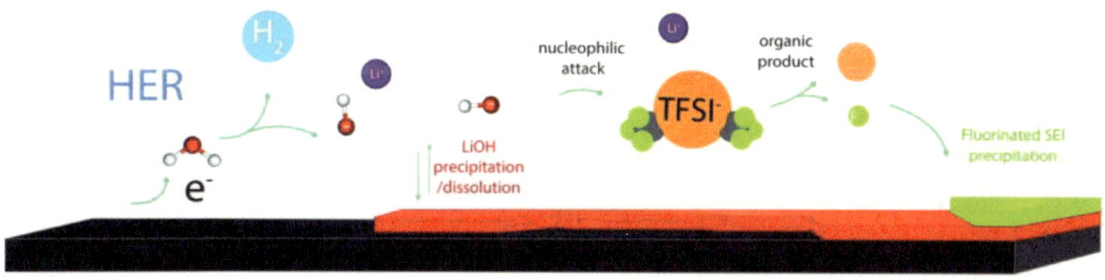

Figure 5.
The diagram of SEI formation. This figure has been reprinted from Ref. [34] with permission from Royal Society of Chemistry.

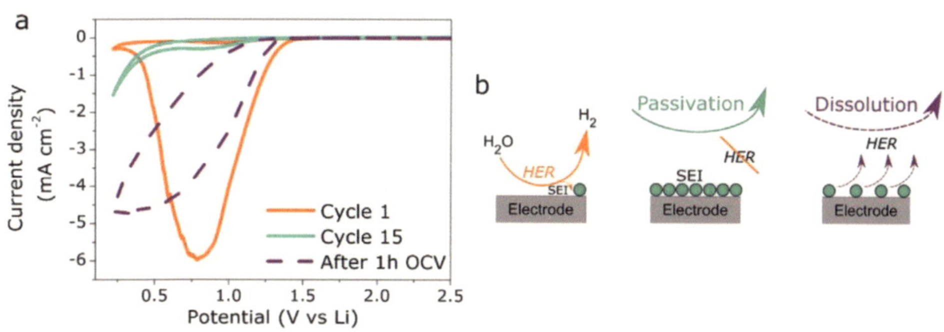

Figure 6.
Assessment of the SEI stability over time in 20 m LiTFSI. a) Cyclic voltammetry at 50 mV s^{-1}, the 1st (red) and 15th (green) cycles, and one cycle after 1 h OCV. b) the illustration of SEI about its partial dissolution after a resting period (1 h). This figure has been reprinted from Ref. [36] with permission from Royal Society of Chemistry.

which means that it requires many cycles (around 15 cycles) to establish the entire SEI on the surface. Moreover, the SEI is not the stable and self-healing layer, meaning that it can be decomposed during cycling and regenerated over a resting period. Thus, future work should be focused on figuring out these issues.

3. Redox-ionic liquid electrolyte

Electrochemical capacitors include the electric double-layer capacitor (EDLC) and pseudocapacitor. EDLC discovers a fast ions aggregation/separation on the interface, resulting in high power but low energy because of the accessible area limit. Pseudocapacitors can share the reversible and fast redox reaction on/near the interface, leading to enhanced energy and moderate power. It motivates us that constructing a new electrolyte, which includes the ionic liquid and redox species, can offer a wide working voltage window and redox element, thereby improving the energy without compromising the power.

3.1 The neat ionic liquid

As the third group of solvents, ionic liquid, compared with water or organic solvents, prevails to achieve a larger voltage window. Moreover, Ionic liquids have a low melting point of less than 100°C due to their low volatility and nature as molten salts [37]. As an enhanced electrolyte, its property can still be stretched, such as high thermal stability, high ionic conductivity, and extensive liquid temperature range. Depended on the above merits, ionic liquid has been intensively researched in energy-storage applications, for example, battery and supercapacitor [38–46].

3.2 The enhanced redox ionic liquid

Although ionic liquid displays sizable merits as an electrolyte, the capacitance/capacity is still far from the expectation. Recently, a promoted electrolyte, by bringing in redox species, is earning intensive attention because it can obtain additional faradaic reactions to boost the capacitance [47–52]. Then, the extra capacitance from redox-active additives can lead to higher energy density. However, it is noted that the redox-active moiety could result in robust self-discharge, and the redox shuttles may cause the degradation of performance, for instance, low coulombic efficiency [53]. It is inspiring that the above challenges can be relieved by redox ionic liquid, which was first introduced by Rochefort et al. [24] in a supercapacitor, where the redox-active moiety was covalently combined with one of ions in ionic liquid. Based on EMIm NTf_2 base, Ferrocene was merged with imidazolium cation or bis(trifluoromethanesulfonyl)imide anion to synthesize EMIm FcNTf and FcEIm NTf_2. As depicted in **Figure 7**, the two-electrode cell with EMIm NTf_2 displayed a rectangular shape, implying a classic double-layer capacitive behavior, namely, occurring ions absorption/desorption on the interface. However, with ferrocene species, the redox peaks could be observed in both EMIm FcNTf and FcEIm NTf_2 electrolytes, which boosted the specific capacitance. Additionally, the Fc-enhanced anion exhibited a strong suppression of self-discharge, resulting in an increase in energy density and stability.

Thanks to the first successful modifying ionic liquid with redox moieties, numerous works have been reported for redox ionic liquid. However, the majority of research has concentrated on integrating a single redox species with a cation or anion, which limits the redox contribution to capacitance. In our team, Fontaine et al. [54] constructed a biredox ionic liquid, as displayed in **Figure 8**, where a

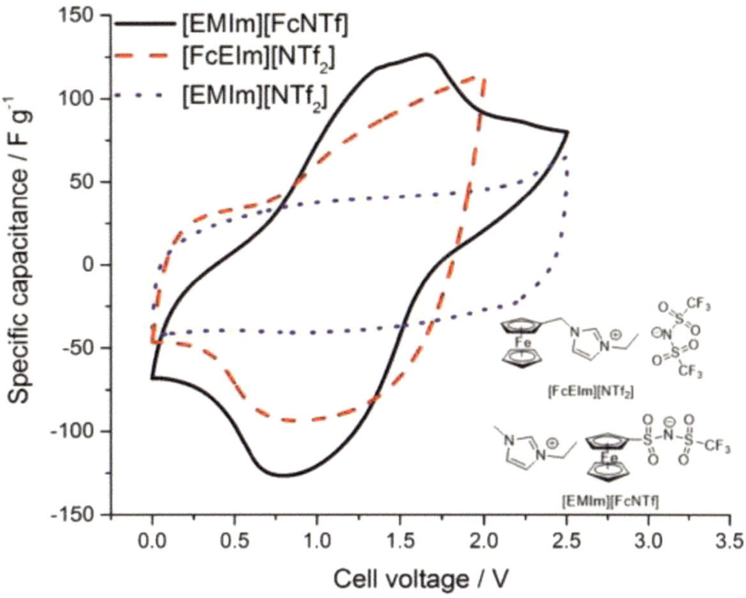

Figure 7.
Cyclic voltammograms based on three ionic liquids. This figure has been reprinted from Ref. [24] with permission from ELSEVIER.

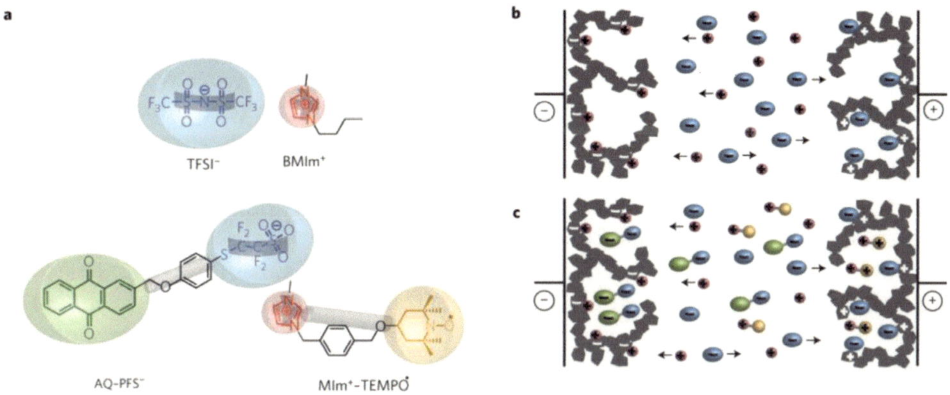

Figure 8.
Structure of BMImTFSI and biredox AQ-PFS⁻ MIm⁺-TEMPO˙. (a) Charge storage mechanism based on ionic liquid (b) and biredox ionic liquid (c). This figure has been reprinted from Ref. [54] with permission from Nature.

perfluorosulfonate anion linked with anthraquinone and a methyl imidazolium cation connected with TEMPO. Over the charge-storage process, in pure BMImTFSI, the corresponding ions separated on the interface and access to the available porous carbon surface following the classic double-layer capacitive concept without involving any Faradaic reaction. However, with the bulky size and high viscosity property, the electro-adsorbed or trapped redox AQ-PFS⁻ and MI⁺-TEMPO˙ were difficult to through the electrode passage, which could suppress the self-discharge. Thus, the trapped biredox species enhanced the capacitance of a few redox reactions, while the large significant capacitance maintained a stable vale over 2000 cycles without obvious degradation. This work, as displayed, sets the stage for constructing a high-energy supercapacitor.

The Biredox ionic liquids are the new paradigm in a supercapacitor, meaning that numerous works are missing to understand the "biredox" mechanism. The central component of understanding the biredox paradigm is a better understanding of electron transfer at the interface, which our team recently proposed [55]. The electrons have to transfer from the electrode to the molecule, then leading to the charge storage in the applied electrolyte. Marcus's theory meticulously describes the transfer process, but the theory must proceed on an ideal flat electrode, where homogeneous and heterogeneous electron transfer occurs. The reality is that it is more complicated with a non-flat electrode, especially, a porous material, as shown in **Figure 9**. According to Marcus's theory, kinetics involves two parameters: the reorganization energy represents the required energy to reorganize the solvent as it approaches the molecule, and the coupling parameter represents the interaction between the molecule and the electrode. Depended on the parameters, it is practical to use Marcus's theory to describe the given case in the pore, where the electrode's porosity is close to the size of the molecule. But, without a good model and experimental basis, it is difficult to depict the effects on electron transfer from the electrode to the molecule. In a word, the related theory is still blank to show the electron transfer movement in the confined pores.

The trick to incorporating redox-active species into ionic liquid is to increase the energy density without sacrificing other properties, such as power density. Except focusing on the electrode materials or upgrading the techniques, the wondrous works about redox and biredox ionic liquid have been successfully reported to boost the energy density with contributions from additional redox moieties, meantime suppressing the in-house self-discharge. By storing charges in parts of electrolytes, this promoted energy density offers a new route to build the enhanced supercapacitor, but the energy is still unsatisfactory as lithium-ion battery. The classic double-layer capacitor's limitation remains that capacitance is limited by the available specific surface area. The redox/biredox species can only migrate to the electrode's accessible surface area, not to the electrode's bulk volume. This means that the number of redox/biredox moieties within the pores is limited, reducing the contribution of redox reactions. The current works are concentrating on double-layer electrode materials, primarily porous carbon. A possible solution is to expand the electrode materials, such as pseudocapacitive electrodes that are still capable of rapid charging. But the specific surface area has marginal effects on the capacitance performance. Then the redox/biredox parts may access to the entire volume of electrodes. However, new issues may exist. The theoretical work should be more explored, especially the charging mechanism on the interface regarding to redox species. The confinement effect with the redox/biredox molecules in the pores may give us new ways to advance our ionic liquid system. The charge storage mechanism on the interface with redox/biredox ionic liquid electrolyte is still open in the supercapacitor.

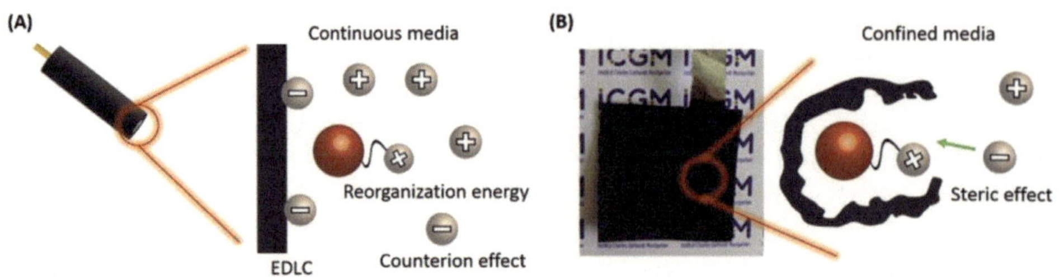

Figure 9.
Electron transfer on the flat electrode (A) and in microporous media (B). This figure has been reprinted from Ref. [55] with permission from ELSEVIER.

4. Conclusion

In this chapter, two advanced electrolytes are profoundly discussed. First, in aqueous system, to boost the energy density, water-in-salt electrolyte is focused to improve the operating working voltage window thanks to the crafted-capably interphase. For traditional aqueous electrolytes, the limited voltage arrange largely hinders the development of advanced supercapacitor device. In the water-in-salt electrolyte, the water can be blocked by the formed interphase, but the ions still are free to pass through. By breaking through the limit of potential, it is not a problem to achieve high energy in aqueous electrolytes, compared with that in organic electrolytes. However, as discussed above, the formed interphase needs many cycles to passivate on the whole electrode surface, and it is not stable enough to keep the long cycling charge–discharge procedure. Moreover, the formation mechanism of solid electrolyte interphase is still not clear. The debate is still focusing on the detailed way to shape the interphase, suggesting that the more theoretical work are required to construct the whole map. For redox-active ionic liquid electrolyte, it is compelling to enhance the capacitance by introducing the extra redox reaction to store more electrons in the electrolyte instead traditional electrode, and meantime the large voltage window is kept. Furthermore, based on designing the redox-enhanced species, the biredox electrolyte can be also obtained to boost the capacitance via using two couples of redox molecules. With matching with the proper size of porosity, a confinement effect occurs, further having a positive impact on improving the capacitance, and then energy density. The defect is that the more suitable model and theoretical research are required to paint the whole picture about the electron transfer from the electrode to the molecule. More significantly, the basic issue is still the available surface area for electric double-layer electrode materials, which is almost reaching the ceiling of its natural property. The redox molecule can still access to the available surface rather than bulk of the volume. Now, it is interesting to use the pseudocapacitive electrode materials to combine with the redox ionic liquid. But it is important that the pseudocapacitive material has to match with the applied redox ionic liquid to exert the bulk volume of electrode instead of the limited surface.

Acknowledgements

The authors thank the French National Research Agency (STORE-EX Labex Project ANR-10-LABX-76-01) and BALWISE (project ID: ANR-19-CE05-0014). O.F also acknowledges the Institut Universitaire de France for the support.

Conflict of interest

The authors declare no conflict of interest.

Author details

Yachao Zhu[1] and Olivier Fontaine[1,2,3*]

1 Institut Charles Gerhardt Montpellier (ICGM), UMR 5253, Université de Montpellier, CNRS, Montpellier Cedex 5, France

2 School of Energy Science and Engineering, Vidyasirimedhi Institute of Science and Technology (VISTEC), Rayong, Thailand

3 Institut Universitaire de France, Paris, France

*Address all correspondence to: olivier.fontaine@umontpellier.fr

References

[1] Y. Gogotsi, P. Simon, Materials for electrochemical capacitors, Nat. Mater. 7 (2008) 845-854.

[2] L.L. Zhang, X.S. Zhao, Carbon-based materials as supercapacitor electrodes, Chem. Soc. Rev. 38 (2009) 2520. https://doi.org/10.1039/b813846j.

[3] Y. Zhu, K. Rajouâ, S. Le Vot, O. Fontaine, P. Simon, F. Favier, Modifications of MXene layers for supercapacitors, Nano Energy. 73 (2020). https://doi.org/10.1016/j.nanoen.2020.104734.

[4] Y. Zhu, W. Chu, N. Wang, T. Lin, W. Yang, J. Wen, X.S. Zhao, Self-assembled Ni/NiO/RGO heterostructures for high-performance supercapacitors, RSC Adv. 5 (2015) 77958-77964. https://doi.org/10.1039/c5ra14790e.

[5] C. Zhong, Y. Deng, W. Hu, J. Qiao, L. Zhang, J. Zhang, A review of electrolyte materials and compositions for electrochemical supercapacitors, Chem. Soc. Rev. 44 (2015) 7484-7539. https://doi.org/10.1039/c5cs00303b.

[6] B. Pal, S. Yang, S. Ramesh, V. Thangadurai, R. Jose, Electrolyte selection for supercapacitive devices: A critical review, Nanoscale Adv. 1 (2019) 3807-3835. https://doi.org/10.1039/c9na00374f.

[7] A. González, E. Goikolea, J.A. Barrena, R. Mysyk, Review on supercapacitors: Technologies and materials, Renew. Sustain. Energy Rev. 58 (2016) 1189-1206. https://doi.org/10.1016/j.rser.2015.12.249.

[8] C. Liu, X. Yan, F. Hu, G. Gao, G. Wu, X. Yang, Toward Superior Capacitive Energy Storage: Recent Advances in Pore Engineering for Dense Electrodes, Adv. Mater. 30 (2018) 1-14. https://doi.org/10.1002/adma.201705713.

[9] S. Bose, T. Kuila, A.K. Mishra, R. Rajasekar, N.H. Kim, J.H. Lee, Carbon-based nanostructured materials and their composites as supercapacitor electrodes, J. Mater. Chem. 22 (2012) 767-784. https://doi.org/10.1039/c1jm14468e.

[10] Y. Wang, Y. Song, Y. Xia, Electrochemical capacitors: Mechanism, materials, systems, characterization and applications, Chem. Soc. Rev. 45 (2016) 5925-5950. https://doi.org/10.1039/c5cs00580a.

[11] Y. Shen, B. Liu, X. Liu, J. Liu, J. Ding, C. Zhong, W. Hu, Water-in-salt electrolyte for safe and high-energy aqueous battery, Energy Storage Mater. 34 (2021) 461-474. https://doi.org/10.1016/j.ensm.2020.10.011.

[12] A.P. Doherty, Redox-active ionic liquids for energy harvesting and storage applications, Curr. Opin. Electrochem. 7 (2018) 61-65. https://doi.org/10.1016/j.coelec.2017.10.009.

[13] L. Lu, T. Xiong, W.S.V. Lee, J. Xue, T.L. Tan, Harmonizing Energy and Power Density toward 2.7 V Asymmetric Aqueous Supercapacitor, Adv. Energy Mater. 8 (2018) 1702630. https://doi.org/10.1002/aenm.201702630.

[14] S. Sathyamoorthi, S. Tubtimkuna, M. Sawangphruk, Influence of structures and functional groups of carbon on working potentials of supercapacitors in neutral aqueous electrolyte: In situ differential electrochemical mass spectrometry, J. Energy Storage. 29 (2020) 101379. https://doi.org/10.1016/j.est.2020.101379.

[15] J. Guo, Y. Ma, K. Zhao, Y. Wang, B. Yang, J. Cui, X. Yan, High-Performance and Ultra-Stable Aqueous Supercapacitors Based on a Green and Low-Cost Water-In-Salt Electrolyte,

ChemElectroChem. 6 (2019) 5433-5438. https://doi.org/10.1002/celc.201901591.

[16] L. Coustan, D. Bélanger, Electrochemical activity of platinum, gold and glassy carbon electrodes in water-in-salt electrolyte, J. Electroanal. Chem. 854 (2019). https://doi.org/10.1016/j.jelechem.2019.113538.

[17] W. Deng, X. Wang, C. Liu, C. Li, J. Chen, N. Zhu, R. Li, M. Xue, Li/K mixed superconcentrated aqueous electrolyte enables high-performance hybrid aqueous supercapacitors, Energy Storage Mater. 20 (2019) 373-379. https://doi.org/10.1016/j.ensm.2018.10.023.

[18] M. Chen, G. Feng, R. Qiao, Water-in-salt electrolytes: An interfacial perspective, Curr. Opin. Colloid Interface Sci. 47 (2020) 99-110. https://doi.org/10.1016/j.cocis.2019.12.011.

[19] M. Zhang, S. Makino, D. Mochizuki, W. Sugimoto, High-performance hybrid supercapacitors enabled by protected lithium negative electrode and "water-in-salt" electrolyte, J. Power Sources. 396 (2018) 498-505. https://doi.org/10.1016/j.jpowsour.2018.06.037.

[20] T. Quan, E. Härk, Y. Xu, I. Ahmet, C. Höhn, S. Mei, Y. Lu, Unveiling the Formation of Solid Electrolyte Interphase and its Temperature Dependence in "water-in-Salt" Supercapacitors, ACS Appl. Mater. Interfaces. (2021). https://doi.org/10.1021/acsami.0c19506.

[21] Y. Kim, M. Hong, H. Oh, Y. Kim, H. Suyama, S. Nakanishi, H.R. Byon, Solid Electrolyte Interphase Revealing Interfacial Electrochemistry on Highly Oriented Pyrolytic Graphite in a Water-in-Salt Electrolyte, J. Phys. Chem. C. 124 (2020) 20135-20142. https://doi.org/10.1021/acs.jpcc.0c05433.

[22] B. Safe, L. Suo, O. Borodin, Y. Wang, X. Rong, W. Sun, X. Fan, S. Xu, M.A. Schroeder, A. V Cresce, F. Wang, C. Yang, Y. Hu, K. Xu, C. Wang, "Water-in-Salt" Electrolyte Makes Aqueous Sodium-Ion, 1701189 (2017) 1-10. https://doi.org/10.1002/aenm.201701189.

[23] Z. Wang, C. Sun, Y. Shi, F. Qi, Q. Wei, X. Li, Z. Sun, B. An, F. Li, A salt-derived solid electrolyte interphase by electroreduction of water-in-salt electrolyte for uniform lithium deposition, J. Power Sources. 439 (2019) 227073. https://doi.org/10.1016/j.jpowsour.2019.227073.

[24] H.J. Xie, B. Gélinas, D. Rochefort, Redox-active electrolyte supercapacitors using electroactive ionic liquids, Electrochem. Commun. 66 (2016) 42-45. https://doi.org/10.1016/j.elecom.2016.02.019.

[25] G. Hernández, M. Işik, D. Mantione, A. Pendashteh, P. Navalpotro, D. Shanmukaraj, R. Marcilla, D. Mecerreyes, Redox-active poly(ionic liquid)s as active materials for energy storage applications, J. Mater. Chem. A. 5 (2017) 16231-16240. https://doi.org/10.1039/c6ta10056b.

[26] R. Balasubramanian, W. Wang, R.W. Murray, Redox ionic liquid phases: Ferrocenated imidazoliums, J. Am. Chem. Soc. 128 (2006) 9994-9995. https://doi.org/10.1021/ja0625327.

[27] T.J. Abraham, D.R. Macfarlane, J.M. Pringle, High Seebeck coefficient redox ionic liquid electrolytes for thermal energy harvesting, Energy Environ. Sci. 6 (2013) 2639-2645. https://doi.org/10.1039/c3ee41608a.

[28] T.Y. Kim, H.W. Lee, M. Stoller, D.R. Dreyer, C.W. Bielawski, R.S. Ruoff, K.S. Suh, High-performance supercapacitors based on poly(ionic liquid)-modified graphene electrodes, ACS Nano. 5 (2011) 436-442. https://doi.org/10.1021/nn101968p.

[29] W.Y. Tsai, R. Lin, S. Murali, L. Li Zhang, J.K. McDonough, R.S. Ruoff, P.L. Taberna, Y. Gogotsi, P. Simon, Outstanding performance of activated graphene based supercapacitors in ionic liquid electrolyte from -50 to 80°C, Nano Energy. 2 (2013) 403-411. https://doi.org/10.1016/j.nanoen.2012.11.006.

[30] Y. Shim, H.J. Kim, Nanoporous carbon supercapacitors in an ionic liquid: A computer simulation study, ACS Nano. 4 (2010) 2345-2355. https://doi.org/10.1021/nn901916m.

[31] C. Arbizzani, M. Biso, D. Cericola, M. Lazzari, F. Soavi, M. Mastragostino, Safe, high-energy supercapacitors based on solvent-free ionic liquid electrolytes, J. Power Sources. 185 (2008) 1575-1579. https://doi.org/10.1016/j.jpowsour.2008.09.016.

[32] N. Yadav, N. Yadav, S.A. Hashmi, Ionic liquid incorporated, redox-active blend polymer electrolyte for high energy density quasi-solid-state carbon supercapacitor, J. Power Sources. 451 (2020) 227771. https://doi.org/10.1016/j.jpowsour.2020.227771.

[33] L. Suo, O. Borodin, T. Gao, M. Olguin, J. Ho, X. Fan, C. Luo, C. Wang, K. Xu, "Water-in-salt" electrolyte enables high-voltage aqueous lithium-ion chemistries, Science (80-.). 350 (2015) 938-943. https://doi.org/10.1126/science.aab1595.

[34] N. Dubouis, P. Lemaire, B. Mirvaux, E. Salager, M. Deschamps, A. Grimaud, The role of the hydrogen evolution reaction in the solid–electrolyte interphase formation mechanism for ""Water-in-Salt"' electrolytes†, Energy Environ. Sci. 11 (2018) 3491-3499. https://doi.org/10.1039/c8ee02456a.

[35] R. Bouchal, Z. Li, C. Bongu, S. Le Vot, R. Berthelot, B. Rotenberg, F. Favier, S.A. Freunberger, M. Salanne, O. Fontaine, Competitive Salt Precipitation/Dissolution During Free-Water Reduction in Water-in-Salt Electrolyte, Angew. Chemie. 132 (2020) 16047-16051. https://doi.org/10.1002/ange.202005378.

[36] L. Droguet, A. Grimaud, O. Fontaine, J. Tarascon, Water-in-Salt Electrolyte (WiSE) for Aqueous Batteries: A Long Way to Practicality, Adv. Energy Mater. 10 (2020) 2002440. https://doi.org/10.1002/aenm.202002440.

[37] A. Balducci, R. Dugas, P.L. Taberna, P. Simon, D. Plée, M. Mastragostino, S. Passerini, High temperature carbon-carbon supercapacitor using ionic liquid as electrolyte, J. Power Sources. 165 (2007) 922-927. https://doi.org/10.1016/j.jpowsour.2006.12.048.

[38] A. Jiang, Z. Wang, Q. Li, M. Dong, Ionic Liquid-Assisted Synthesis of Hierarchical One-Dimensional MoP/NPC for High-Performance Supercapacitor and Electrocatalysis, ACS Sustain. Chem. Eng. 8 (2020) 6343-6351. https://doi.org/10.1021/acssuschemeng.0c00238.

[39] H. Zhou, C. Liu, J.C. Wu, M. Liu, D. Zhang, H. Song, X. Zhang, H. Gao, J. Yang, D. Chen, Boosting the electrochemical performance through proton transfer for the Zn-ion hybrid supercapacitor with both ionic liquid and organic electrolytes, J. Mater. Chem. A. 7 (2019) 9708-9715. https://doi.org/10.1039/c9ta01256g.

[40] L. Miao, H. Duan, Z. Wang, Y. Lv, W. Xiong, D. Zhu, L. Gan, L. Li, M. Liu, Improving the pore-ion size compatibility between poly(ionic liquid)-derived carbons and high-voltage electrolytes for high energy-power supercapacitors, Chem. Eng. J. 382 (2020) 122945. https://doi.org/10.1016/j.cej.2019.122945.

[41] D. Wang, Y. Wang, H. Liu, W. Xu, L. Xu, Unusual carbon nanomesh constructed by interconnected carbon

nanocages for ionic liquid-based supercapacitor with superior rate capability, Chem. Eng. J. 342 (2018) 474-483. https://doi.org/10.1016/j.cej.2018.02.085.

[42] F. Poli, D. Momodu, G. Emanuele, A. Terella, B.K. Mutuma, M. Letizia, N. Manyala, F. Soavi, Electrochimica Acta Pullulan-ionic liquid-based supercapacitor : A novel , smart combination of components for an easy-to-dispose device, Electrochim. Acta. 338 (2020) 135872. https://doi.org/10.1016/j.electacta.2020.135872.

[43] D.A. Rakov, F. Chen, S.A. Ferdousi, H. Li, T. Pathirana, A.N. Simonov, P.C. Howlett, R. Atkin, M. Forsyth, Engineering high-energy-density sodium battery anodes for improved cycling with superconcentrated ionic-liquid electrolytes, Nat. Mater. 19 (2020) 4-10. https://doi.org/10.1038/s41563-020-0673-0.

[44] U. Pal, F. Chen, D. Gyabang, T. Pathirana, B. Roy, R. Kerr, D.R. Macfarlane, M. Armand, P.C. Howlett, M. Forsyth, concentrated ionic liquid electrolyte for long-life practical lithium metal battery applications †, (2020) 18826-18839. https://doi.org/10.1039/d0ta06344d.

[45] A.K. Tripathi, Ionic liquid e based solid electrolytes (ionogels) for application in rechargeable lithium battery, Mater. Today Energy. 20 (2021) 100643. https://doi.org/10.1016/j.mtener.2021.100643.

[46] H. Gupta, R. Kumar, High-Voltage Nickel-Rich NMC Cathode Material with Ionic- Liquid-Based Polymer Electrolytes for Rechargeable Lithium-Metal Batteries **, (2020) 3597-3605. https://doi.org/10.1002/celc.202000608.

[47] L. Yang, K. Zhuo, X. Xu, Z. Zhang, Q. Du, Y. Chen, D. Sun, J. Wang, Redox-active phthalocyanine-decorated graphene aerogels for high-performance supercapacitors based on ionic liquid electrolyte, J. Mater. Chem. A. 8 (2020) 21789-21796. https://doi.org/10.1039/d0ta08054c.

[48] K. Manickavasakam, S. Suresh Balaji, S. Kaipannan, A.G. Karthick Raj, S. Veeman, S. Marappan, Electrochemical Performance of Thespesia Populnea Seeds Derived Activated Carbon - Supercapacitor and Its Improved Specific Energy in Redox Additive Electrolytes, J. Energy Storage. 32 (2020) 101939. https://doi.org/10.1016/j.est.2020.101939.

[49] L.Q. Fan, Q.M. Tu, C.L. Geng, Y.L. Wang, S.J. Sun, Y.F. Huang, J.H. Wu, Improved redox-active ionic liquid-based ionogel electrolyte by introducing carbon nanotubes for application in all-solid-state supercapacitors, Int. J. Hydrogen Energy. 45 (2020) 17131-17139. https://doi.org/10.1016/j.ijhydene.2020.04.193.

[50] Y. Wang, C. Malveau, D. Rochefort, Solid-state NMR and electrochemical dilatometry study of charge storage in supercapacitor with redox ionic liquid electrolyte, Energy Storage Mater. 20 (2019) 80-88. https://doi.org/10.1016/j.ensm.2019.03.023.

[51] N. Ma, N. Phattharasupakun, J. Wutthiprom, M. Sawangphruk, Electrochimica Acta High-performance hybrid supercapacitor of mixed-valence manganese oxide / N-doped graphene aerogel nano fl ower using an ionic liquid with a redox additive as the electrolyte : In situ electrochemical X-ray absorption spectroscopy, Electrochim. Acta. 271 (2018) 110-119. https://doi.org/10.1016/j.electacta.2018.03.116.

[52] E. Cevik, A. Bozkurt, M. Hassan, M.A. Gondal, T.F. Qahtan, Redox-Mediated Poly(2-acrylamido-2-methyl-1-propanesulfonic acid)/Ammonium Molybdate Hydrogels for Highly Effective Flexible Supercapacitors,

ChemElectroChem. 6 (2019) 2876-2882. https://doi.org/10.1002/celc.201900490.

[53] C. Bodin, C. Sekhar Bongur, M. Deschanels, S. Catrouillet, S. Le Vot, F. Favier, O. Fontaine, Shuttle Effect Quantification for Redox Ionic Liquid Electrolyte Correlated to the Coulombic Efficiency of Supercapacitors, Batter. Supercaps. 3 (2020) 1193-1200. https://doi.org/10.1002/batt.202000084.

[54] C. Bodin, E. Mourad, D. Zigah, S. Le Vot, S.A. Freunberger, F. Favier, O. Fontaine, Biredox ionic liquids: New opportunities toward high performance supercapacitors, Faraday Discuss. 206 (2018) 393-404. https://doi.org/10.1039/c7fd00174f.

[55] O. Fontaine, A deeper understanding of the electron transfer is the key to the success of biredox ionic liquids, Energy Storage Mater. 21 (2019) 240-245. https://doi.org/10.1016/j.ensm.2019.06.023.

Section 8

Wearable Supercapacitors

Chapter 11

Wearable Supercapacitors, Performance, and Future Trends

Litty V. Thekkekara and Imtiaz Ahmed Khan

Abstract

The progress in portable technologies demands compactable energy harvesting and storage. In recent years, carbon-based lightweight and wearable supercapacitors are the new energy storage trends in the market. Moreover, the non-volatile nature, long durability, eco-friendliness, and electrostatic interaction mechanism of supercapacitors make it a better choice than traditional batteries. This chapter will focus on the progress of the wearable supercapacitor developments, the preferred material, design choices for energy storage, and their performance. We will be discussing the integrability of these supercapacitors with the next generation wearable technologies like sensors for health monitoring, biosensing and e-textiles. Besides, we will investigate the limitations and challenges involves in realizing those supercapacitor integrated technologies.

Keywords: carbon, energy storages, E-textiles, health monitoring, biosensing

1. Introduction

Wearable technologies can be defined as intelligent devices that can be worn or attached to the skin's surface where they can detect, analyze and transfer the data to relevant systems [1]. The best example of E-textiles in the current market is sensors integrated apparels incorporating technologies like antennas, energy harvesters, and sensors [2]. These can enable the development of next-generation self-reliant wearable applications for wireless communications [3, 4], health sensing and monitoring [5–7], and light-emitting devices [8–10], which can find applications in smart cities, remote areas, telecommunications, and biomedical industries [11].

In the current commercial market, we can find more than 1000 types of wearables. Some of them include products from famous brands like Apple smartwatches which includes healthcare monitoring, fitness trackers from Garmin, integrated sensors in apparels from Nike and Adidas [12]. In the modern days, the research is focused on developing textiles itself as a sensor to monitor the body functions [13]. The expected market size of wearables is around $57,653 million by 2022 [14].

In general, energy harvesters like piezo generators, which utilizes the energy delivered from the mechanical motions within the body functions, or solar cells which harvest energy from the Sun are utilized as a medium of energy harvesters, and traditional batteries are used as energy storage in wearable devices [15–18]. Further, wireless charging is a promising concept for the powering of wearables [19]. However, current wearable technologies are limited for continuous monitoring due to the power failures from the integrated coin-cells or pouch cell lithium-ion batteries [20]. Besides, batteries are volatile and suffer from heating issues [21].

Flexible supercapacitors are an alternative energy storage to be considered for wearable technologies due to the features like fast charging nature, long durability, integrability with the technologies, and eco-friendliness [22–24]. In the chapter, we will discuss how supercapacitors can be used as energy storage for supporting wearable technologies and the challenges involved in it.

2. Performance

Supercapacitors are generally divided into two types: *symmetric and asymmetric* based on their working mechanism. Symmetric supercapacitors can store electrical energy through ion adsorption–desorption (electrical double layer capacitive, EDLC) mechanism. In contrast, asymmetric supercapacitors work through the Faradaic reaction (pseudocapacitive) between the electrode and electrolyte [25]. Besides, there are two types of configuration for the supercapacitors: (1) those have a vertical structure comprising a separator sandwiched between two electrodes are known as sandwich supercapacitors and (2) a micro supercapacitor (MSC) is based on an interdigitated structures in the same plane as the current collectors, separated from each other by an insulating gap (typically in the range of 10–100 s of micrometers) [26]. Finally, an electrolyte is coated on top and between the electrodes and ensures ion transport along the basal plane of the electrodes [27].

In MSC, the insulating gap between the electrodes and electrode width decides the migrating distance of electrolyte ions, which generates the equivalent series resistance (ESR) [28]. With the reduction of the ESR, the performance of the MSCs can be improved, and the calculation can be. The electrode thickness is another factor deciding the storage capacity of the MSCs [29]. The ESR energy storage capacity of the supercapacitors is defined through the energy density (E_{den}), and the rate of the charge transfer process, power density (P_{den}) is analyzed using measurements like cyclic voltammetry (CV), galvanic charge–discharge (CC), and impedance measurements [27].

The formulas for calculating the ESR, E_{den} (Wh cm^{-2}) and Pden (W cm^{-2}), can be defined as follows;

$$R_{ESR} = V_{drop} / 2i \tag{1}$$

$$E_{den} = C_v \times (\Delta E)^2 \times (2 \times 3600) \tag{2}$$

$$P_{den} = (\Delta E)^2 / (4ESR \times V) \tag{3}$$

where R_{ESR} is the internal voltage drop at the beginning of the discharge, V_{drop}, at a constant current density, i calculated from the CC measurements, C_v is the volumetric capacitance, ΔE is the operating voltage window in Volts, V is the volume of the electrodes.

3. Materials and designs

Materials and designs are an essential factor that decides the energy storage performance like flexibility, lightweight, storage capacity, how fast electrolyte ions can move within the device, and electrochemical window of the device [27, 30].

For wearables, a particular type of supercapacitors needs to be designed to match the above specific requirements. In this session, we will discuss these aspects in detail.

3.1 Materials

3.1.1 Electrodes

The electrodes of supercapacitors require high surface area, long term stability, resistance to electrochemical oxidation or reduction, the capability of multiple cycling materials, optimum pore size distribution, minimized ohmic resistance with the contacts, sufficient electrode-electrolyte solution contact interface, mechanical integrity, and less self-discharge [25, 27, 31].

EDLCs mainly utilize carbon-based materials for the electrodes due to their high performance [32]. One of the first EDLC developed employs activated carbon as the electrode material, which exhibited capacity values of 2 F cm^{-2} in H_2S0_4 solutions [33]. However, carbon exhibits slow oxidation, besides having high ESR. The low performance of carbon is due to poor particle to particle contact of the agglomerates as well as the high ionic resistance from the electrolyte distributed in the microporous structure, resulting in the poor high-frequency characteristics of carbon-based capacitors. On the other hand, the carbon nanotubes do not produce satisfactory capacitance unless a conducting polymer [34] is used to form a pseudocapacitance.

Graphene is a form of carbon with a high surface area up to 2675 m^2 g^{-1} and intrinsic capacitance of 21 μF cm^{-2}, which set the upper limit of EDLC capacitance of all carbon-based materials [35]. Besides, both faces of graphene sheets are readily accessible by the electrolyte. However, in practical applications, the surface area of graphene will be much reduced due to agglomeration. Graphene-related materials like reduced graphene oxide are cost-effective and widely used electrode materials for EDLCs [36].

Pseudocapacitors which are asymmetric supercapacitors using different materials like RuO_2, Manganese oxide (MnO_2), and conductive polymers like polyaniline (PANI) with or without the symmetric electrode materials, becomes a direction of interest to achieve the high-performance supercapacitors [37]. For example, the hybrid of ultrathin supercapacitors made of MnO2 sheets and graphene sheets using the direct laser writing method offers an electrochemically active surface for fast absorption/desorption electrolyte ions (22). The contributions of additional interfaces at the hybridized interlayer areas to accelerate charge transport during the charge/discharge process resulting in an energy density and power density of 2.4 mWh cm^{-3} and 298 mW cm^{-3}, respectively. Flexible supercapacitors based on manganese hexacyanoferrate-manganese oxide and electrochemically reduced graphene oxide electrode materials (MnHCF-MnO*x*/ErGO) exhibiting a remarkable areal capacitance of 16.8 mF cm^{-2} and considerable energy and power density of 0.5 mWh cm^{-2} and 0.0023 mW cm^{-2} [38].

Another approach is to use metals like the well-connected nanoporous gold film to fabricate interdigital electrode materials for supercapacitors with high mechanical flexibility [39]. These supercapacitors exhibit a capacitance of 127 F cm^{-3} and an energy density of 0.045 Wh cm^{-3}. The gold metal is known for its high electrical conductivity, and the concept adopted can be effectively used to integrate with devices in a lesser aerial footprint.

3.1.2 Electrolytes

The electrolyte of supercapacitors has a crucial role in deciding properties such as the energy density, power density, internal resistance, rate performance, operating temperature range, cycling lifetime, self-discharge, non-volatile nature, and toxicity of the energy storage. The electrochemical range of an electrolyte decides the cell voltage window of the energy storage like the batteries and supercapacitors [25] and is governed by the equation,

$$E = \frac{1}{2} C V^2 \tag{4}$$

where E = energy density, C = specific capacitance and V = cell voltage.

The electrolytes used in energy storage can be classified as liquid electrolytes and solid/quasi-solid state electrolytes [40]. Liquid electrolytes can be further classified as aqueous electrolytes with a voltage range of 1.0 to 1.3 V, organic electrolytes within the voltage range of 1 to 2 V, and ionic liquids (IL) with a voltage range of 3.5 to 4.0 V [41]. The solid or quasi-solid state electrolytes can be classified as organic and inorganic electrolytes with a voltage range of 2.5 to 2.7 V.

Among different electrolytes, aqueous-based electrolytes possess high conductivity and capacitance. However, they are limited by low cell voltage windows whereas organic, and IL electrolytes can operate at higher cell voltage windows. ILs are used in wearable energy storage owing to their interesting properties like non-flammability, low vapor pressure, and significant operating potential window. Solid-state electrolytes are devoid of leakage issues but are limited by the low conductivity [42].

3.1.3 Designs

It is highly recommended to have an optimized design for supercapacitor electrodes for high output performance. In the commercial supercapacitors, sandwich structures in which electrode-electrolyte-electrode configuration is utilized [26]. Nevertheless, these designs can result in bulky storages, which is less favorable for lightweight wearable technologies.

On the other hand, the printed supercapacitor, based on 2D planar interelectrode configuration, is utilized as the basic designs for the printed electrodes. However, the performance is limited in comparison to the sandwich counterpart [43]. This has lead to the consideration of other designs like a spiral, split rings, and onion petals which demonstrated an increase in the electrode-electrolyte interactions in the supercapacitor [44]. A considerable enhancement in the storage capacitance and power density was offered with the utilization of the fractals designs [18], which can offer an unlock towards the development of high capacity miniaturized [23, 45, 46] as well as large scale supercapacitors that can be integrated with the textiles and other wearables [47].

The other area of recent interest in the designs is the origami concepts to improve the performance of the printed supercapacitors [48]. The research used the active materials in suitable geometry to create self-folding structures to perform folding or unfolding functions without having kinetic movements due to external forces.

3.1.4 Encapsulation

The wearable energy storage will be exposed to conditions like water moisture conditions from sweating, washing, weather conditions, and atmospheric pollution. All these conditions can adversely affect the performance of these energy storage. Besides, the presence of corrosive and volatile electrolytes can be dangerous to the user's health. Effective encapsulation is an essential condition to sustain a safer storage performance while maintaining a flexible nature [49–51]. Ecoflex and polyvinyl alcohol (PVA) are the standard encapsulations used for the current wearable energy storage [47, 52].

4. Types of wearable supercapacitors

4.1 Coin/pouch supercapacitors

The most commonly used energy storages for portable devices like smartwatches are the coin and pouch cells. Due to the well-developed production line for the long-used methodology, there is high interest in the adaptability of the technique with the extension towards the supercapacitors for the emerging wearable technologies [53–55]. Pooachi et al. recently developed prototypes of coin cells and pouch cells from nitrogen-doped reduced graphene oxide electrodes with phenylenediamine-mediated organic electrolyte with a high specific capacitance of 563 and 340 F g^{-1} with high energy density 149.4 and 77.2 Wh·kg^{-1} at 1 A g^{-1} [56].

4.2 Printed supercapacitors

A less complex solution to the energy storage demands of wearables, printed technologies offer highly adaptive methodologies for producing adhesion between the electrodes and current collectors/substrates and eliminates the requirement of inert additives for active materials. In general, supercapacitor printing techniques can be classified into two categories, techniques which do not require a template (example-inkjet printing and 3D printing) and techniques utilizing a template (example-screen printing). All these techniques must be coordinated with the printable materials for improving electrochemical and mechanical performance in a less footprint [57].

Among them, laser-induced graphene based supercapacitors attained exceptional attention due to the cost-effectiveness and ability to integrate with wearable applications in specific scales [58, 59]. Recently, demonstrated textile integrated solar graphene energy storages of 100 cm^2 with a performance of an areal capacitance, 49 mF cm^{-2}, energy density, 6.73 mWh cm^{-2}, power density, 2.5 mW cm^{-2}, and stretchability up to 200%, which can effectively be utilized for the realization of functional textiles to support applications like sensors, and displays [47, 60].

Printing high-performance supercapacitors in lesser footprints can develop three-dimensional (3D) supercapacitors [61]. The concept of the layer by layer stacking of individual supercapacitors obtained from the laser-induced graphenes from PET sheets which result in an areal capacitance >9 mF cm^{-2} [62, 63]. This methodology is a promising direction towards the future of energy storages to be considered for the ultra-portable and flexible applications. Besides, there are reports of using multilayered structures made of rGO/Au particles [64]. The development of additional features like stretchability up to 50% in 1000 stretch cycles with the 3D supercapacitors will contribute towards the withstanding of the deformations and prevention against the performance degradation [65].

Nonimpact printing technology like inkjet printing is an additive-based approach that can create patterns either continuously or in steps by propelling droplets of liquid precursor materials onto various substrates without the aid of predesigned masks through the control of printer head and ink toner [66]. The formation of printed features depends on the capability of the inkjet printing apparatus, the viscosity, surface tension, dispersibility of the inks, and the wettability of the substrates to be printed on [67]. Yu and co-workers reported paper-based, all-solid, flexible, planar supercapacitors by inkjet printing PEDOT: PSS-CNT/silver nanoparticle as the electrode material [68]. The obtained microdevices were able to demonstrate rate capability up to 10 000 mV s^{-1}, fast frequency response (relaxation time constant of 8.5 ms), high volumetric specific capacitance (23.6 F cm^{-3}), and long cycle stability (92% capacitance retention after 10, 000 cycles).

Screen printing is another approach for printing and is conducted through squeegee by pressing the ink down with enough force to penetrate through prepatterned masks (screen or stencil) onto the desired substrate [69]. The process can be conducted on both rigid (silicon or glass) and flexible substrates (textiles or papers) and can reach a minimum resolution of 30–50 μm [70]. The quality of the resulting features depends on the stenciling techniques, the printability of the inks, and the affinity between the ink and substrate. Even though this method is capable of mass production, some issues like printable ink must have a high viscosity, and suitable shear-thinning property limits the potential. Using this methodology, Lu et al. prepared an all-solid MSC by screen-printing $FeOOH/MnO_2$ composites on different substrates like PET, paper, and textile. The fully printed supercapacitor exhibits a high area-specific capacitance of 5.7 mF cm^{-2} and an energy density of 0.0005 mWh cm^{-2} at a power density of 0.04 mW cm^{-2} [71].

Roll to roll printing of supercapacitors through the low-temperature laser annealing process of roll-to-roll (R2R) printed metal nanoparticle (NP) ink on a polymer substrate is an area of interest that can have the largescale commercial applications [72]. Another approach is to laser-print the toner on metal foils, followed by thermal annealing in a hydrogen environment, finally resulting in the patterned thin graphitic carbon or graphene electrodes for supercapacitors. The electrochemical cells are made of graphene–graphitic carbon electrodes, which can be roll-to-roll printed [73].

4.3 Yarn based supercapacitors

E-textiles are the new frontiers in the history of fabrics that incorporate the technologies like displays and sensing [60, 74, 75]. Textiles or fabrics are flexible materials, which use fibers originating from natural or synthetic as fundamental building blocks, with a considerable length to diameter ratio (~1000 to 1) [76]. The fabrics can be in the form of staples or filament.

Natural fibers such as cotton staple fiber have a limited length. Filament fiber tends to be continuous in length, whereas silk is an example of a natural filament An example of synthetic fibers are filament fibers. Both staple fibers and threads can be made into yarns and fabrics. A more recent form of textile is electrospun, where the fabrication is achieved by applying a high voltage to an aqueous polymeric solution. The polymer can be organic or inorganic, depending on the intended application. *Yarn* is a long strand constituting textile fibers, filaments, or similar material converted to a textile fabric through weaving, knitting, or other intertwining techniques [77]. The yarn can be formed from various manufacturing processes. The spinning process is used to convert fibers into yarns, usually by twisting together several fibers. Usually, in spinning, fibers are aligned in one axis and are axially twisted. The direction of twist is denoted by 'S' or 'Z' twist.

When looking for 2D fabric devices such as supercapacitor devices, the usual architecture found in the sandwich [78] and in-plane (planar) configurations [79]. The in-plane architecture is more flexible than its sandwich structure counterpart due to its lightweight and flexibility (structural design), making it a more suitable option when working with 2D active electrode materials [80]. In the case of 1D yarn supercapacitors, common existing device shapes include coaxial [81], core-spun [82], parallel [83], two-ply twisted [84], and helical structures [85].

5. Future trends—scalability and integration of wearable supercapacitors

The major hindrance of the current wearable energy storage to be commercialized is its limitation for large-scale fabrications to meet the power requirements of integrated technologies [86]. Besides, for textile-based wearable technologies, another challenge is developing a single textile unit incorporating energy storage and devices together. The achievement of cost-effective all-in-one wearable technologies without compensating the performance of the technologies is a great challenge in this area [87–96]. Some fo the research groups demonstrated the laboratory protoypes in this area and in **Figure 1**, we summarized these studies.

Only a few groups so far that reported about the possibilities of developing their energy storages can be developed into self-powered wearable technologies using the industrial machinery because of the low performance [97]. Besides, the cost-effectiveness of the process needs to be attractive in comparison to the existing battery technology. Nevertheless, the development of printing methodologies like screen printing [56], inkjet printing [61], and laser printing seems to be a solution to the stitching issues currently faced by the yarn-based energy storages. The successful generation of high-performance flexible, lightweight, miniaturized supercapacitors will enable a step closer to realizing eco-friendly, non-volatile portable and wearable devices.

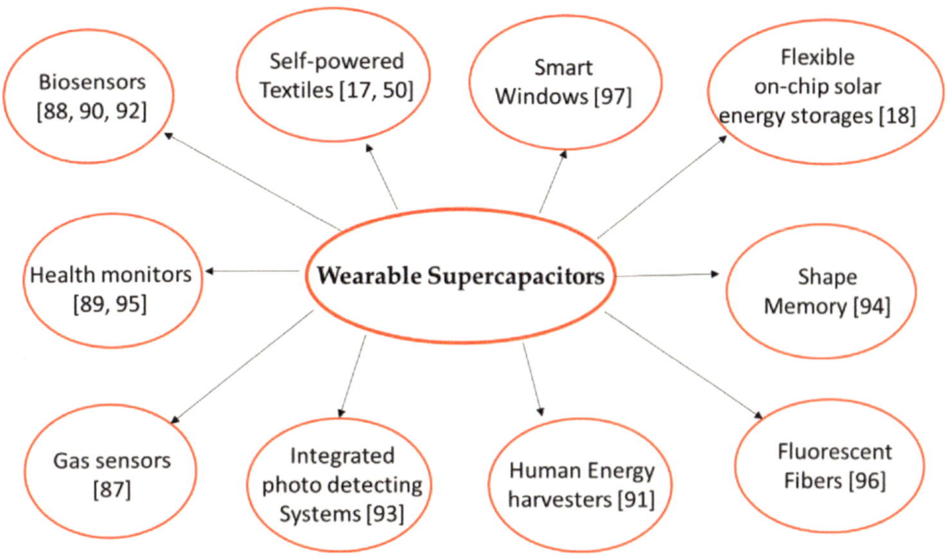

Figure 1.
Applications benefited from the wearable supercapacitors.

6. Conclusions

Supercapacitors which leaves the further possibilities of miniaturization without compromising the high energy storage capacity and transfer rate provides the scope of improvement to be adopted as an eco-friendly, non-voltaile energy storage source for the future wearable technologies.

Acknowledgements

LVT acknowledge the RMIT seed fund, 2018 to support this work. IAK thanks Australian Research Conucil (ARC) linkage grant- LP 110100455 for the PhD scholarship.

Conflict of interest

The authors declare no conflict-of-interest declaration.

Appendices and nomenclature

MSC	Microsupercapacitor
EDLC	Electrochemical double layer capacitance

Author details

Litty V. Thekkekara[1,2*] and Imtiaz Ahmed Khan[1]

1 Department of Manufacturing, Materials and Mechatronics, School of Engineering, RMIT University, Melbourne, Australia

2 Functional Materials and Microsystems, Department of Electronic and Telecommunications, School of Engineering, RMIT University, Melbourne, Australia

*Address all correspondence to: littyvarghese.thekkekara@rmit.edu.au

References

[1] Schwab, K., The Fourth Industrial Revolution. 2017: Penguin Books Limited.

[2] Schneegass, S. and O. Amft, Smart textiles. 2017: Springer.

[3] Huang, X., et al., Highly flexible and conductive printed graphene for wireless wearable communications applications. Scientific reports, 2015. 5(1): p. 1-8.

[4] Rein, M., et al., Diode fibres for fabric-based optical communications. Nature, 2018. 560(7717): p. 214-218.

[5] Rodgers, M.M., V.M. Pai, and R.S. Conroy, Recent advances in wearable sensors for health monitoring. IEEE Sensors Journal, 2014. 15(6): p. 3119-3126.

[6] He, W., et al., Integrated textile sensor patch for real-time and multiplex sweat analysis. Science Advances, 2019. 5(11): p. eaax0649.

[7] Van Langenhove, L., Smart Textiles for Medicine and Healthcare: Materials, Systems and Applications. 2007: Elsevier.

[8] Choi, S., et al., Highly flexible and efficient fabric-based organic light-emitting devices for clothing-shaped wearable displays. Scientific reports, 2017. 7(1): p. 1-8.

[9] Graham-Rowe, D., Photonic fabrics take shape. Nature Photonics, 2007. 1(1): p. 6-7.

[10] Zhang, Z., et al., Textile display for electronic and brain-interfaced communications. Advanced Materials, 2018. 30(18): p. 1800323.

[11] Shi, J., et al., Smart textile-integrated microelectronic Systems for Wearable Applications. Advanced Materials, 2020. 32(5): p. 1901958.

[12] Seneviratne, S., et al., A survey of wearable devices and challenges. IEEE Communications Surveys & Tutorials, 2017. 19(4): p. 2573-2620.

[13] Koncar, V., Smart Textiles and their Applications. 2016: Woodhead Publishing.

[14] Research and Markets Offers Report: Global Wearable Technology Market Analysis & Trends - Industry Forecast to 2025. Telecommunications Reports, Wireless News, 2018.

[15] Wang, X., et al., Flexible energy-storage devices: Design consideration and recent Progress. Advanced Materials, 2014. 26(28): p. 4763-4782.

[16] Ostfeld, A.E., et al., High-performance flexible energy storage and harvesting system for wearable electronics. Scientific reports, 2016. 6(1): p. 1-10.

[17] Wen, Z., et al., Self-powered textile for wearable electronics by hybridizing fiber-shaped nanogenerators, solar cells, and supercapacitors. Science Advances, 2016. 2(10): p. e1600097.

[18] Thekkekara, L.V. and M. Gu, Bioinspired fractal electrodes for solar energy storages. Scientific reports, 2017. 7(1): p. 1-9.

[19] Burket, C., Wireless charging opportunities and challenges for wearables: One size does not fit all. IEEE Power Electronics Magazine, 2017. 4(4): p. 53-57.

[20] Çiçek, M., Wearable technologies and its future applications. International Journal of Electrical, Electronics and Data Communication, 2015. 3(4): p. 45-50.

[21] Habibipour, A., A.M. Padyab, and A. Ståhlbröst. Social, ethical and

ecological issues in wearable technologies. in AMCIS 2019, Twenty-Fifth Americas Conference on Information Systems, Cancun, México, Augusti 15-17 2019. 2019.

[22] Hashmi, S., Supercapacitor: An emerging power source. National Academy Science Letters, 2004. 27(1-2): p. 27-46.

[23] Thekkekara, L.V., X. Chen, and M. Gu, Two-photon-induced stretchable graphene supercapacitors. Scientific Reports, 2018. 8(1): p. 11722.

[24] Miller, J.R. and P. Simon, Electrochemical capacitors for energy management. Science, 2008. 321(5889): p. 651-652.

[25] Conway, B., Springer Science & Business Media. 1999.

[26] Yu, A., V. Chabot, and J. Zhang, Electrochemical Supercapacitors for Energy Storage and Delivery: Fundamentals and Applications. 2013: Taylor & Francis.

[27] Conway, B.E., Electrochemical Supercapacitors: Scientific Fundamentals and Technological Applications. 2013: Springer Science & Business Media.

[28] Liu, N. and Y. Gao, Recent progress in micro-supercapacitors with in-plane interdigital electrode architecture. Small, 2017. 13(45): p. 1701989.

[29] Xiong, G., et al., A review of graphene-based electrochemical microsupercapacitors. Electroanalysis, 2014. 26(1): p. 30-51.

[30] Simon, P. and Y. Gogotsi, Materials for electrochemical capacitors. Nanoscience and technology: a collection of reviews from Nature journals, 2010: p. 320-329.

[31] Stoller, M.D. and R.S. Ruoff, Best practice methods for determining an electrode material's performance for ultracapacitors. Energy & Environmental Science, 2010. 3(9): p. 1294-1301.

[32] Inagaki, M., H. Konno, and O. Tanaike, Carbon materials for electrochemical capacitors. Journal of power sources, 2010. 195(24): p. 7880-7903.

[33] Boos, D.L., Electrolytic capacitor having carbon paste electrodes. 1970, Google Patents.

[34] Obreja, V.V., On the performance of supercapacitors with electrodes based on carbon nanotubes and carbon activated material—A review. Physica E: Low-dimensional Systems and Nanostructures, 2008. 40(7): p. 2596-2605.

[35] Simon, P. and Y. Gogotsi, Materials for electrochemical capacitors. Nature materials, 2008. 7(11): p. 845-854.

[36] Purkait, T., et al., High-performance flexible supercapacitors based on electrochemically tailored three-dimensional reduced graphene oxide networks. Scientific reports, 2018. 8(1): p. 1-13.

[37] Wang, J., et al., Pseudocapacitive materials for electrochemical capacitors: From rational synthesis to capacitance optimization. National Science Review, 2017. 4(1): p. 71-90.

[38] Liang, J., et al., All-printed MnHCF-MnOx-based high-performance flexible supercapacitors. Advanced Energy Materials, 2020. 10(12): p. 2000022.

[39] Zhang, C., et al., Planar integration of flexible micro-supercapacitors with ultrafast charge and discharge based on interdigital nanoporous gold electrodes on a chip. Journal of Materials Chemistry A, 2016. 4(24): p. 9502-9510.

[40] Zhong, C., et al., A review of electrolyte materials and compositions

for electrochemical supercapacitors. Chemical Society Reviews, 2015. 44(21): p. 7484-7539.

[41] Armand, M., et al., Ionic-liquid materials for the electrochemical challenges of the future. Nature materials, 2009. 8(8): p. 621-629.

[42] Samui, A.B. and P. Sivaraman, 11 - Solid Polymer Electrolytes for Supercapacitors, in Polymer Electrolytes. 2010, Woodhead Publishing. p. 431-470.

[43] Wang, F., et al., Latest advances in supercapacitors: From new electrode materials to novel device designs. Chemical Society Reviews, 2017. 46(22): p. 6816-6854.

[44] Gao, W., et al., Direct laser writing of micro-supercapacitors on hydrated graphite oxide films. Nature nanotechnology, 2011. 6(8): p. 496-500.

[45] Soavi, F., et al., Miniaturized supercapacitors: Key materials and structures towards autonomous and sustainable devices and systems. Journal of power sources, 2016. 326: p. 717-725.

[46] Kyeremateng, N.A., T. Brousse, and D. Pech, Microsupercapacitors as miniaturized energy-storage components for on-chip electronics. Nature nanotechnology, 2017. 12(1): p. 7-15.

[47] Thekkekara, L.V. and M. Gu, Large-scale waterproof and stretchable textile-integrated laser-printed graphene energy storages. Scientific reports, 2019. 9(1): p. 1-7.

[48] Nam, I., et al., All-solid-state, origami-type foldable supercapacitor chips with integrated series circuit analogues. Energy & Environmental Science, 2014. 7(3): p. 1095-1102.

[49] Ji, W., et al., Polypyrrole encapsulation on flower-like porous NiO for advanced high-performance supercapacitors. Chemical Communications, 2015. 51(36): p. 7669-7672.

[50] Kim, H., et al., Encapsulated, high-performance, stretchable array of stacked planar micro-supercapacitors as waterproof wearable energy storage devices. ACS applied materials & interfaces, 2016. 8(25): p. 16016-16025.

[51] Kavinkumar, T., et al., Interface-modulated uniform outer nanolayer: A category of electrodes of nanolayer-encapsulated core-shell configuration for supercapacitors. Nano Energy, 2021. 81: p. 105667.

[52] Li, H. and J. Liang, Recent development of printed micro-supercapacitors: Printable materials, printing technologies, and perspectives. Advanced Materials, 2020. 32(3): p. 1805864.

[53] Hung, T.-F., et al., High-mass loading hierarchically porous activated carbon electrode for pouch-type supercapacitors with propylene carbonate-based electrolyte. Nanomaterials, 2021. 11(3): p. 785.

[54] Ojha, M., et al., Holey graphitic carbon nano-flakes with enhanced storage characteristics scaled to a pouch cell supercapacitor. Fuel, 2021. 285: p. 119246.

[55] Roberts, A.J., et al. Supercapacitors: From coin cell to 800 F pouch cell. in ECS Meeting Abstracts. 2016. IOP Publishing.

[56] Poochai, C., et al., High performance coin-cell and pouch-cell supercapacitors based on nitrogen-doped reduced graphene oxide electrodes with phenylenediamine-mediated organic electrolyte. Applied Surface Science, 2019. 489: p. 989-1001.

[57] Gopi, C.V.M., et al., Recent progress of advanced energy storage materials for

flexible and wearable supercapacitor: From design and development to applications. Journal of Energy Storage, 2020. 27: p. 101035.

[58] Fu, X.-Y., et al., Laser fabrication of graphene-based supercapacitors. Photonics Research, 2020. 8(4): p. 577-588.

[59] Maher, F.E., et al., Laser scribing of high-performance and flexible graphene-based electrochemical capacitors, Science. 2012. p. 1326-1330.

[60] Ergoktas, M.S., et al., Graphene-enabled adaptive infrared textiles. Nano Letters, 2020. 20(7): p. 5346-5352.

[61] Zhao, C., et al., Three dimensional (3D) printed electrodes for interdigitated supercapacitors. Electrochemistry Communications, 2014. 41: p. 20-23.

[62] Lin, J., et al., Laser-induced porous graphene films from commercial polymers. Nature communications, 2014. 5(1): p. 1-8.

[63] Peng, Z., et al., Flexible and stackable laser-induced graphene supercapacitors. ACS applied materials & interfaces, 2015. 7(5): p. 3414-3419.

[64] Li, R.-Z., et al., High-rate in-plane micro-supercapacitors scribed onto photo paper using in situ femtolaser-reduced graphene oxide/Au nanoparticle microelectrodes. Energy & Environmental Science, 2016. 9(4): p. 1458-1467.

[65] Li, X., et al., 3D-printed stretchable micro-supercapacitor with remarkable areal performance. Advanced Energy Materials, 2020. 10(14): p. 1903794.

[66] De Gans, B.J., P.C. Duineveld, and U.S. Schubert, Inkjet printing of polymers: State of the art and future developments. Advanced materials, 2004. 16(3): p. 203-213.

[67] Wu L., et al., Emerging progress of inkjet technology in printing optical materials. Advanced Optical Materials, 2016. 4(12): p. 1915-1932.

[68] Liu, W., et al., Based all-solid-state flexible micro-supercapacitors with ultra-high rate and rapid frequency response capabilities. Journal of Materials Chemistry A, 2016. 4(10): p. 3754-3764.

[69] Xu, Y., et al., Screen-printable thin film supercapacitor device utilizing graphene/polyaniline inks. Advanced Energy Materials, 2013. 3(8): p. 1035-1040.

[70] Cao, X., et al., Screen printing as a scalable and low-cost approach for rigid and flexible thin-film transistors using separated carbon nanotubes. ACS nano, 2014. 8(12): p. 12769-12776.

[71] Lu, Q., et al., Facile synthesis of amorphous FeOOH/MnO2 composites as screen-printed electrode materials for all-printed solid-state flexible supercapacitors. Journal of Power Sources, 2017. 361: p. 31-38.

[72] Yeo, J., et al., Flexible supercapacitor fabrication by room temperature rapid laser processing of roll-to-roll printed metal nanoparticle ink for wearable electronics application. Journal of Power Sources, 2014. 246: p. 562-568.

[73] Kang, S., et al., Roll-to-roll laser-printed graphene–graphitic carbon electrodes for high-performance supercapacitors. ACS applied materials & interfaces, 2018. 10(1): p. 1033-1038.

[74] Service, R.F., Electronic textiles charge ahead. Science, 2003. 301(5635): p. 909-911.

[75] Dias, T., Electronic Textiles: Smart Fabrics and Wearable Technology. 2015: Woodhead Publishing.

[76] Murthy, H.S., Introduction to Textile Fibres. 2016: CRC Press.

[77] ASTM, D123-17 Standard Terminology Related to Textiles. 2017: West Conshohocken, PA.

[78] Hu, L., et al., Stretchable, porous, and conductive energy textiles. Nano Letters, 2010. 10(2): p. 708-714.

[79] Raj, C.J., et al., Highly flexible and planar supercapacitors using graphite flakes/polypyrrole in polymer lapping film. ACS applied materials & interfaces, 2015. 7(24): p. 13405-13414.

[80] Peng, X., et al., Two dimensional nanomaterials for flexible supercapacitors. Chemical Society Reviews, 2014. 43(10): p. 3303-3323.

[81] Harrison, D., et al., A coaxial single fibre supercapacitor for energy storage. Physical Chemistry Chemical Physics, 2013. 15(29): p. 12215-12219.

[82] Zhang, D., et al., Core-spun carbon nanotube yarn supercapacitors for wearable electronic textiles. Acs Nano, 2014. 8(5): p. 4571-4579.

[83] Cao, Y., et al., Boosting supercapacitor performance of carbon fibres using electrochemically reduced graphene oxide additives. Physical Chemistry Chemical Physics, 2013. 15(45): p. 19550-19556.

[84] Chen, Q., et al., MnO 2-modified hierarchical graphene fiber electrochemical supercapacitor. Journal of Power Sources, 2014. 247: p. 32-39.

[85] Bae, J., et al., Fiber supercapacitors made of nanowire-fiber hybrid structures for wearable/flexible energy storage. Angewandte Chemie International Edition, 2011. 50(7): p. 1683-1687.

[86] Tsai, I.-L., et al., Graphene-enhanced electrodes for scalable supercapacitors. Electrochimica Acta, 2017. 257: p. 372-379.

[87] Li, L., et al., Flexible planar concentric circular micro-supercapacitor arrays for wearable gas sensing application. Nano Energy, 2017. 41: p. 261-268.

[88] Park, H., et al., Dynamically stretchable supercapacitor for powering an integrated biosensor in an all-in-one textile system. ACS nano, 2019. 13(9): p. 10469-10480.

[89] Lu, Y., et al., Wearable sweat monitoring system with integrated micro-supercapacitors. Nano Energy, 2019. 58: p. 624-632.

[90] Sun, Tongrui, et al., Wearable textile supercapacitors for self-powered enzyme-free Smartsensors. ACS applied materials & interfaces, 2020. 12(19): 21779-21787.

[91] He, Z, et al. Piezoelectric-driven self-powered patterned electrochromic supercapacitor for human motion energy harvesting. ACS Sustainable Chemistry & Engineering, 2018. 7 (1): 1745-1752.

[92] Vaghasiya, Jayraj V., et al. Integrated biomonitoring sensing with wearable asymmetric supercapacitors based on Ti3C2 MXene and 1T-phase WS2 Nanosheets. Advanced Functional Materials, 2020. 30(39): 2003673.

[93] Wang, X, et al. Fiber-based flexible all-solid-state asymmetric supercapacitors for integrated photodetecting system. Angewandte Chemie, 2014. 126(7): 1880-1884.

[94] Huang, Yang, et al. A shape memory supercapacitor and its application in smart energy storage textiles. Journal of Materials Chemistry A, 2016 4(4): 1290-1297.

[95] Huang, J, et al. Self-powered integrated system of a strain sensor and flexible all-solid-state supercapacitor by using a high performance ionic

organohydrogel. Materials Horizons, 2020 7(8): 2085-2096.

[96] Liao, M, et al. Multicolor, fluorescent supercapacitor fiber. Small, 2018 14(43): 1702052.

[97] Wang, K., et al., Integrated energy storage and electrochromic function in one flexible device: An energy storage smart window. Energy & Environmental Science, 2012. 5(8): p. 8384-8389.

Section 9

Application for Energy Harvesting

Chapter 12

Piezoelectric-Driven Self-Powered Supercapacitor for Wearable Device Applications

A. Bharathi Sankar Ammaiyappan and Seyezhai Ramalingam

Abstract

Supercapacitors are the most promising energy storage devices that bridge the gap between capacitors and batteries. They can reach energy density close to the batteries and power density to the conventional capacitors. Several kinds of research have been carried out in the field of supercapacitors for the development of promising electrode and electrolyte materials as well as device fabrications to breakthrough in energy storage systems with diverse applications in electronics. They have a broad range of applications as they can deliver a huge power within a very short time. The applications of supercapacitors in several sectors like consumer and portable electronics, transportation and vehicles, power backup, biomedical, military, aerospace, etc.

Keywords: Piezoelectric, Self-powered, Supercapacitor, Energy harvesting

1. Introduction

1.1 Supercapacitor as next generation energy storage device

The continuous development of smart electronic devices and industrial high-power devices has made a strong requirement of high-performance electrochemical energy storage devices. Out of various electrochemical energy storage devices available in the market. Lithium-ion batteries (LIBs) and supercapacitors (SCs) find a considerable application. However, SC device offers a wide variety of benefits in the electronic devices. Some of the benefits related to the SC device are it offers low-cost material E.g., Activated Carbon (AC), lightweight and long durability cell performance for the smart and portable electronic devices. In the past few decades, the rapid growth in the advancements in the development of portable electronic systems has stimulated the research interest among researchers to design and develop innovative electrical energy storage devices with maximum energy density and power density. SCs or supercapacitors are electrochemical energy storage device that store electrical charges via electrochemical and electrostatic charge storage mechanism takes place at the surface of the electrode. SCs hold high energy performance as compared to common dielectric capacitors and hence SCs are extensively utilized not only for powering several portable electronic devices but also plug-in hybrid electric vehicles.

SCs showed outstanding potential as an energy storage device as compared to LIB systems because of their high-power density, super-longer cycle life, and longer cyclic stability. SCs are already being used worldwide in various applications ranging from automotive, renewable energy to consumer electronics and even employed in the bio-medical field of application. SCs are steadily paving the way for hybrid power storage applications, such as complimentary batteries, especially in two-wheeler applications. Various market reports estimate the global demand for SCs to grow tremendously, primarily driven by different consumer electronics and automotive applications, to provide backup power. SCs provide the necessary power backup required for the smooth functioning of applications such as video calling, high-quality image output which is captured during the night-time using cameras, wireless communications, and global positioning system (GPS) navigation, etc.

As per the IDTech Ex report, the global supercapacitor market is expected to reach US$ 8.3 billion by 2025. The market is projected to grow at a compound annual growth rate (CAGR) of 30 percent till 2025 as depicted in **Figure 1**. Consumer electronics and automotive will be the highest revenue-generating segments during this period. Although the supercapacitor market is at a nascent stage, ample emerging growth opportunities are available for the future. Highlighting features such as regenerative braking and easy application in hybrid electric vehicles (EVs) are making supercapacitors the best-suited device in automotive applications. Increasing demand for renewable energy generation has been observed in countries across Europe, Asia, and the USA, which would further fuel the supercapacitor market growth.

Interestingly, the supercapacitors market in India is projected to grow at a compound annual growth rate (CAGR) of around 16% during the year between 2012 and 2022 on the account of a huge demand for supercapacitors from the consumer electronics segment (**Figure 1**). Supercapacitors are used in several devices in the consumer electronics category such as smartphones, laptops, TVs, cameras, lighting appliances, GPS devices, etc. Moreover, the evolution of supercapacitors as a sustainable energy storage solution, growth of the electric vehicles market, and increasing capacities of supercapacitors resulting in their application in the wind and solar power sectors is anticipated to boost demand for supercapacitors in India during the next five years.

According to 'Supercapacitor Market Landscape Study' prepared by the Electronic Industries Association of India (ELCINA), an apex industry association supporting the electronics and IT hardware manufacturing industry, by 2020, the

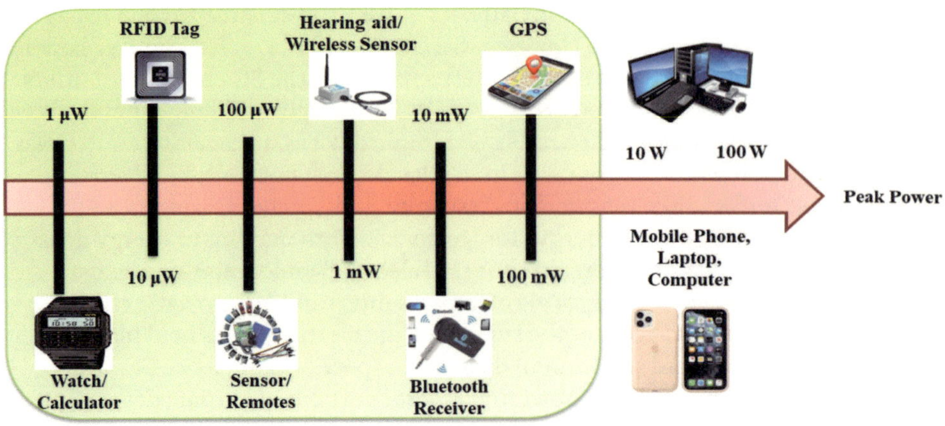

Figure 1.
Schematic diagram of energy harvesting various power range.

overall market opportunity for supercapacitors in India will be approximately 1 billion units, with major demand from consumer electronics, EVs, renewable energy, railway, and defense, etc. Supercapacitor technology has not yet been commercialized in the Indian market, and supercapacitors are not yet common in use. Only a handful of companies and research laboratories are working on their manufacturing in the lab proto-type level. However, considering growing concerns of environmental issues, global warming, air pollution, green energy as well as government initiatives "Make in India" in which India becoming a global manufacturing hub, the Indian market is poised for an explosive demand for supercapacitors in various sectors due to their diverse applications.

1.2 Supercapacitor developers

The requirement for the SC device is increasing day by day. So there is a potential opportunity for the SC developers in the market. Recently, the number of manufactures and developers for the SC device is increasing rapidly. The two categories of SCs device is available in the market. i.e. aqueous and non-aqueous-based SC. However, the non-aqueous-based SC developers have occupied the majority of the market. The main reason is organic electrolyte offers high operating voltage. So it offers high energy density without sacrificing its power density. The solvent mostly used in organic SCs is acetonitrile (ACN) and propylene carbonate (PC). The two solvents ACN and PC are responsible for the high voltage operation of the SC device. However, currently, 50% of the SC manufacturer's available in the market offers non-toxic and non-combustible electrolyte with unique performance and quality. Based on the Electrochemical energy storage requirements the application of SC varies. The various available SC manufacturers and developers will be discussed in detail in the following sections.

In general, SC manufacturers offer to build the device in a cylindrical or prismatic design. In cylindrical manufacturing, electrode cells are deposited onto a sheet and wound like a jelly roll into a cylinder. A casing maintains the capacitor's shape. Cylindrical designs are more common, but more prismatic designs are evolving so that EDLC based supercapacitors can substitute for LIBs in consumer electronics. Most established companies in the SC area are based in Japan, however many startups are located in the USA and Europe.

1.3 Global supercapacitor and supercapacitor manufacturers

Maxwell technologies are the United States of America (USA) based SC company/manufacture. The company was founded in the year of 1965. But now, Maxwell Technologies is one of the leading supercapacitor manufacturers in the global supercapacitor market with is having huge customers around the globe through its device performance. Maxwell's supercapacitors are rapidly evolving and increasingly used technology that can charge and discharge energy quickly and efficiently. Maxwell technology Supercapacitors supplement the main source of energy, that cannot usually cause rapid explosions, such as a combustion engine, battery, or fuel cell. The horizon of the future looks bright for supercapacitors, that are readily distinguished as a potential alternative energy source. The various products of Maxwell technology offer for commercial applications, including high voltage capacitors, microelectronic components, and systems, telecommunication equipment as environmentally safe backup power supplies.

Second, Murata Manufacturing is a Japan-based supercapacitor company that was laid in the year of 1944. Murata is also involved in the design and development of Supercapacitor for the global market. Murata is mainly involved in the manufacture of lower farad capacitors, and sales of electronic parts, including components for supercapacitors and piezoelectric products. Nanoramic Laboratories, formerly known as FastCAP Systems, is a USA-based SC manufacture that specializes in nanocarbon materials and high operating temperature SC. FastCAP Supercapacitors, specializes in harsh environment energy storage systems, producing supercapacitors capable of operating in temperatures up to 150°C and under conditions of high shock and vibration.

To find a replacement for the Lithium-ion batteries in small scale, Paper Battery Company (PBC Tech), is the manufacturers for the ultrathin supercapacitors as replacements for lithium batteries. The PBC tech's SC offers ultrathin supercapacitors of 1 Farad. Also, the company targets the consumer electronics, wearables, and wireless sensor markets.

On the other hand, Skeleton Technologies is the SC manufacture from Estonia who manufactures and develops EDLCs with high energy and power density. The company serves many applications to the market includes transportation, automotive, industrial, and renewable energy markets. Also, skeleton technology SC offers high-end carbon and adsorption materials to the energy storage application. Like the same, Yunasko is a Ukraine based SC manufacture laid in the year of 2010, who develops prismatic type EDLCs capacitor to the market. Its R&D facility is located in Ukraine. Yunasko EDLC cells are manufactured in special prismatic encasements made from multi-layered laminated aluminum foil. Depending on the number of cells connected in series, the module can handle the voltage range from 2.7 V (single-cell) up to 750 V (large assemblies of cells).

2. Energy harvesting for self-powered wearable device applications

Piezoelectric harvesting has proven to be a new solution to replace lead acid batteries and lithium-ion batteries in remote power supply applications. Unfortunately, low energy efficiency is shown in **Figure 1** to make effective use of energy enhancement in daily life. The Power Electric perspective was introduced after reviewing the current research literature on piezoelectric energy harvesting focusing on low frequency vibration mechanics from circuit design. Piezoelectric energy can be used for low power generation due to its high energy storage capacity. The energy harvester to convert stored crop energy includes electrical energy and renewable energy. The reason for choosing piezoelectric is the high energy storage capacity as shown in **Figure 2**. In piezoelectric energy harvesting, the piezoelectric sensor is used as a harvesting vibrational material and the supercapacitor storage element [1–4].

Recent advances in super-low electronic microcontrollers have produced Device that provide unprecedented integration with the amount of energy used. These chip systems have powerful energy-saving schemes, such as shutting down idle work. In fact, these devices require very little power to use, so most sensors are wireless because they run easily on batteries. Unfortunately, the batteries need to be replaced regularly, which is very expensive to produce and the most difficult maintenance [5–7]. The sensory atmosphere is an effective wireless energy solution for harvesting the surrounding mechanical, thermal, solar and electro-magnetic forces in **Figure 2**.

Analog devices provide ICs with high power for super-low power generation applications. Power management products that convert energy from vibratory (piezoelectric), photovoltaic (solar) and thermal (TEC, TEG, thermopiles, thermocouples) provide high-performance conversion to source controlled voltage and charge batteries, supercapacitors and storage devices. Booster converters operate on

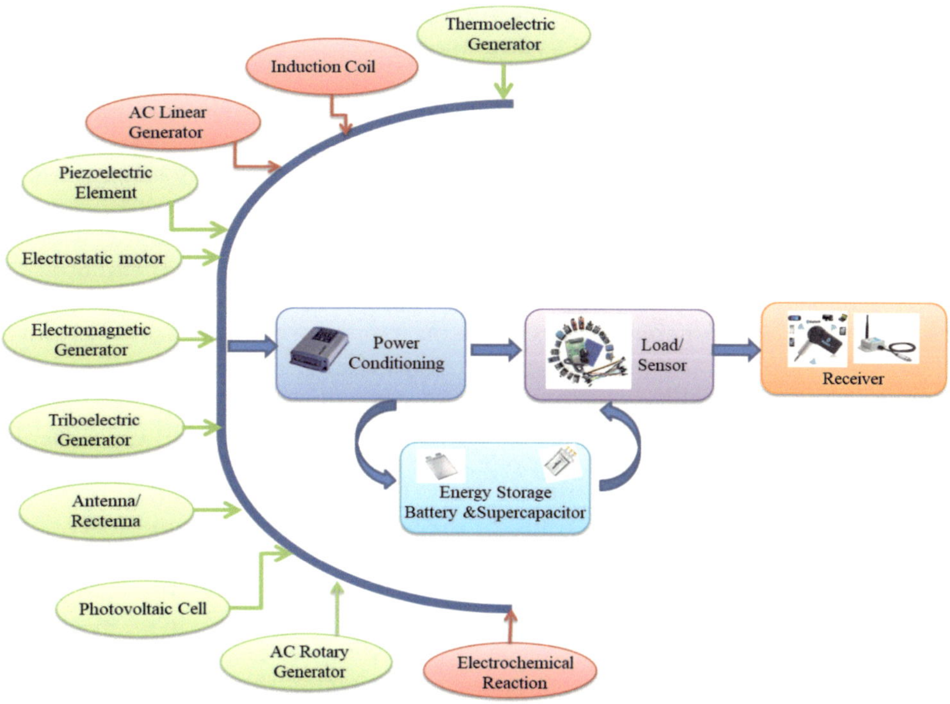

Figure 2.
Schematic diagram of ambient energy source and energy harvesting technologies.

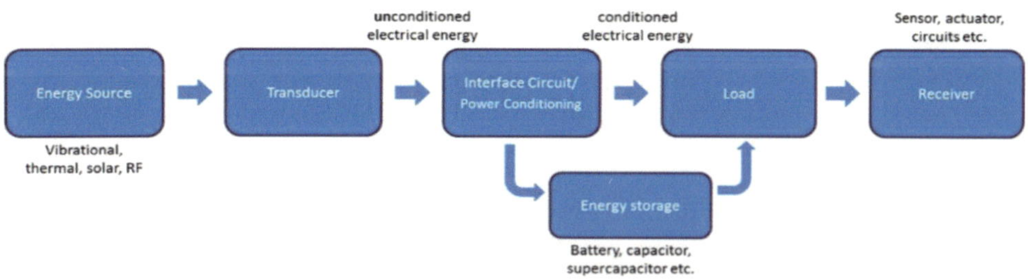

Figure 3.
Block diagram of energy harvesting system.

small 20 mV high capacity battery chargers, expanding the possibilities of automation and industrial control, wireless sensors, navigation, applications and portable electronics [8–10]. Current low power controllers, app amps, comparators, voltage monitors, ADCs, DACs and low power storage devices provide additional blocks for standalone systems as shown in **Figure 3**.

The renewable energy environment is very energy efficient, so energy harvesters are a good source of energy for IoT applications, eliminating the need to replace and dispose of batteries. However, low-power harvesters are often unable to provide the high capacity needed to collect and transfer data. This article shows you how to use a supercapacitor mounted on an energy harvester to provide the maximum power needed using a small piezoelectric strip as a case study. A standard power supply consists of a power harvester that connects a large charging circuit directly to the supercapacitor with a direct load. The high C and low ESR of the supercapacitor keep the electrical energy constant for the load while breaking its high power [11–14].

Over the past decade, flexible electronics, wearable electronics, and portable devices have become of great importance in various fields such as mobility, consumer technology, biomedical, sports, clean and natural energy. Therefore, since high power consumption is required, these electronics require smart energy storage devices. In various energy storage systems, supercapacitors are an important device that can deliver high power in a very short time with new energy storage methods. Supercapacitors are a good energy saving device that closes the gap between electrolytic capacitors and batteries. Compared to traditional electrolytic capacitors, the supercapacitor can use batteries and power supplies near the power supply. Much research has been done in the field of supercapacitors in the manufacture of fine electrodes and electrolyte materials and in the development of energy-efficient electronic storage systems. Initially, supercapacitors were produced from high-carbon materials by forming a double-layer structure. They have excellent power with very high discharge rates and long cycle life. However, over time many types of pseudo supercapacitor gain interest due to the high energy levels obtained from their faradaic process as a result of the surface redox reaction. Recently, hybrids of two building materials have been improved using the low energy content of carbon material and poor performance, as well as the interaction of the two components to overcome the surface area of pseudo supercapacitors Therefore, it is possible to achieve efficiency between capacitors and conventional batteries by improving power and maximum power. In the first stage, supercapacitors were traditionally used as hybrid electric vehicles as well as alternative energy-saving devices in batteries or fuel cells. Since then, this attractive electrical device has slowly evolved into a backup power provider with batteries and fuel cells. Nowadays, supercapacitors occupy applications in many fields such as wearable electronics, flexible electronics, portable electronics, electric vehicle transportation, electric power storage, biomedical, military and aerospace. This paper details the multiple uses of supercapacitors, their real-time application in modern technology and the end of convenience trends [15–24].

3. Methodology

Ambient power sources include light, temperature variation, vibration beam, RF signal signals or other sources that can generate electricity through a transducer. For example, small solar panels have been using hand-held electronic devices for years and can produce 100 mW / cm2 in direct sunlight and 100 μW /cm2 in indirect light. See beck devices convert thermal energy into electrical energy, where the temperature gradient is obtained. Energy sources vary according to body temperature, with earth producing 10 $\mu W/cm^2$ for a fire stack capable of producing 10 mW/cm^2. Piezoelectric elements can produce up to 100 $\mu W/cm^2$ depending on their size and formulation and are shown in **Figure 4**. RF energy is collected by the harvesting antenna and emits 100 pW/cm^2.

Piezoelectric Transducer Application **Figure 4** illustrates the piezoelectric system that, when plugged into an air transmission, produces 100 μwatts of power at 3.3 V. The deviation of the piezoelectric element is 0.5 cm at a frequency of 50 Hz. Harvesting of mechanical energy in human motion is an attractive way to obtain clean and stable electrical energy. Piezoelectricity is the electrical energy generated by mechanical pressure (e.g. walking, running). When pressure is applied to an object, there is a negative charge on the elongated side and a positive charge on the suppressed crystal piezoelectric side. When

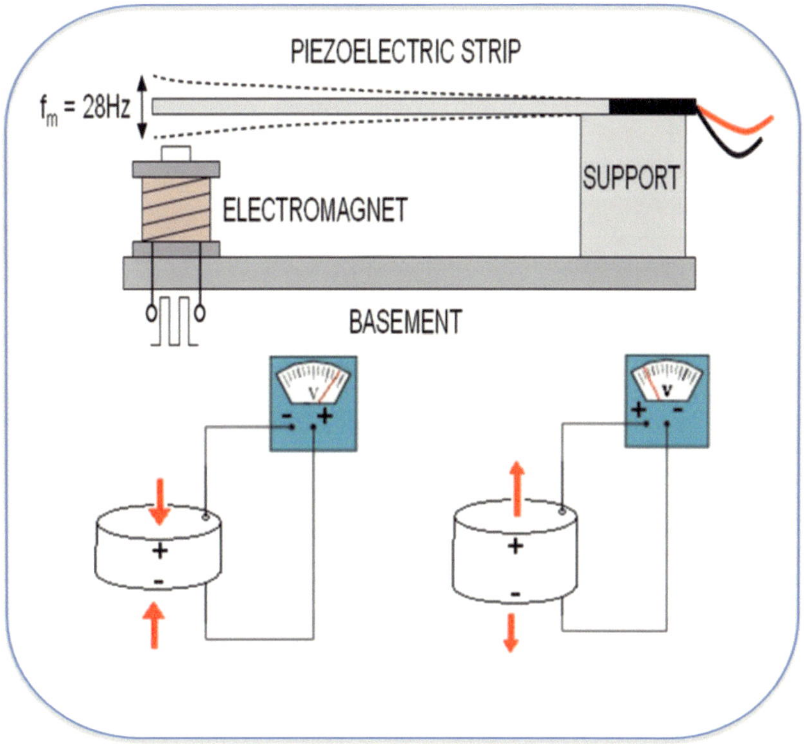

Figure 4.
Block diagram of principle of piezoelectric materials.

pressure is applied, electrical energy flows through the whole. Widely used resources: solar power, triboelectric nano generator power and piezoelectric power. This study focuses on piezoelectricity because it relies on mechanical stresses to obtain electrical energy and some resources are not always reliable. Compared to the three nano generators for maximum energy storage capacity, piezoelectric has more energy saving properties than other energy generation methods.

Piezoelectric sensors should be placed on the two main parts of the shoe where high pressure will occur. A piezoelectric generator is installed inside the shoe insole. The shoe has two points, where the pressure is very strong and the heel and toe, and the piezoelectric sensor wing are the exact location shown in **Figure 5**. The piezoelectric array arrangement fixes the shoe insole. One sensor can produce 3–5 volts in a constant pressure application, in which case four sensors are connected in parallel, which increases the probability of obtaining the highest gains. It is more advantageous to use piezo polymeric materials than piezoelectric materials when it comes to using sensors, as polymer films can be easily made in different sizes. However piezo ceramic sensor was used in this work because it is available for sale at low cost.

The design includes series-connected piezoelectric generator units. The front panel features piezoelectric generators with a straight forward arrangement and a rear panel with circular rotation. The acquisition and charging side collects intermittent or continuous power input from the piezoelectric generator and charges its power properly in the supercapacitor bank. During the charging process, the supercapacitor voltage is constantly monitored. When it reaches 5.2 V, it can power the module's output rectifier and charging circuit.

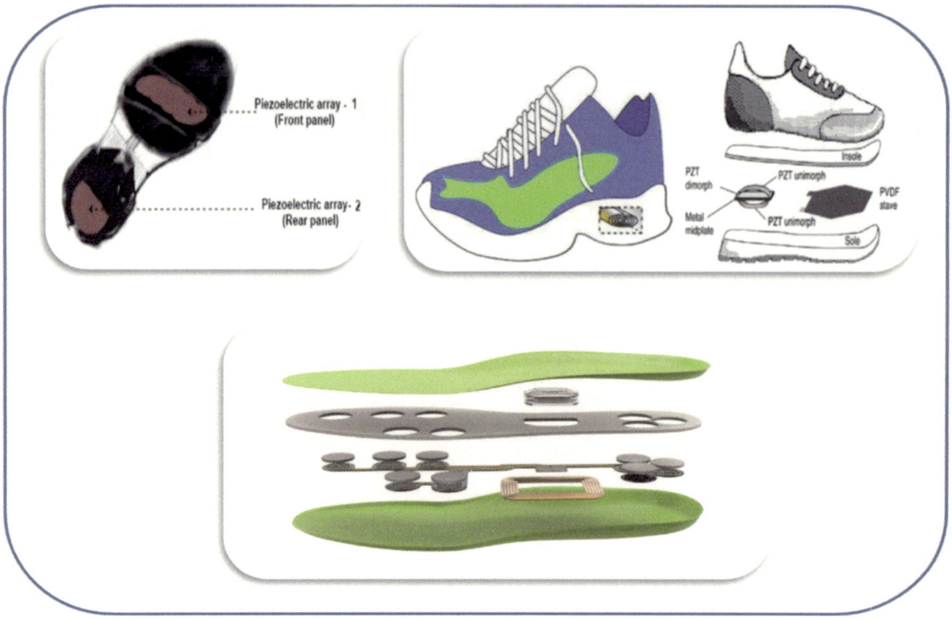

Figure 5.
Arrangement of piezoelectric generator inside a shoe insole.

The piezoelectric generator is mounted on the shoe. As a person moves, pressure builds up on the ground and this pressure can be converted into electrical energy and used to charge the supercapacitor. This energy storage system uses biomedical sensor applications.

There are many studies that successfully explore power augmentation in laboratories, but the overall efficiency of the underlying systems is limited to the trade-off between the capacities of each system. For example, some researchers pay full attention to increasing the productivity of a piezoelectric source, while dissipating the controller's energy reduces the useful energy stored in the energy buffer. The functional cycle of a supercapacitor with a high-performance life cycle is investigated based on a systematic analysis of piezoelectric power generation from a power management perspective.

4. Results

Successful design of complete wireless systems requires energy-saving microcontrollers and transducers used from low-power electronic applications. Since both are now readily available, the active power conversion product capable of changing the output of the transducer is active energy. The LTC3588–1 shown in **Figure 6** is a complete power generation solution designed for high impedance sources such as piezoelectric transducers. It features a full bridge rectifier and a highly efficient buck converter of low damage bridge rectifier that transfers power from the device to the output at a controlled capacity capable of supporting loads up to 100 mA. The LTC3588–1 is available in a 10-lead MSE package with 3 mm × 3 mm DFN.

Figure 7 shows the power generation system with the methods of harvester, transducer and power condition circuit that converts this stored electrical energy into a controlled power supply. The transgender may also need a

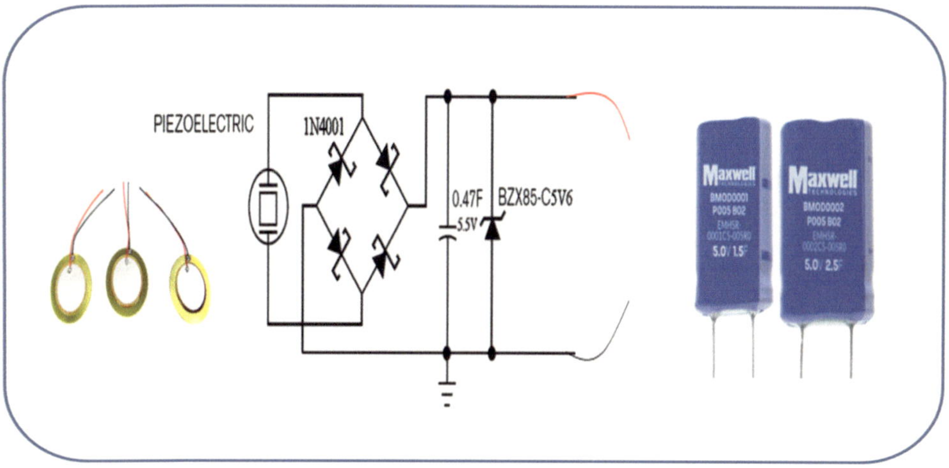

Figure 6.
Simple circuit diagram of piezoelectric-driven charging supercapacitor.

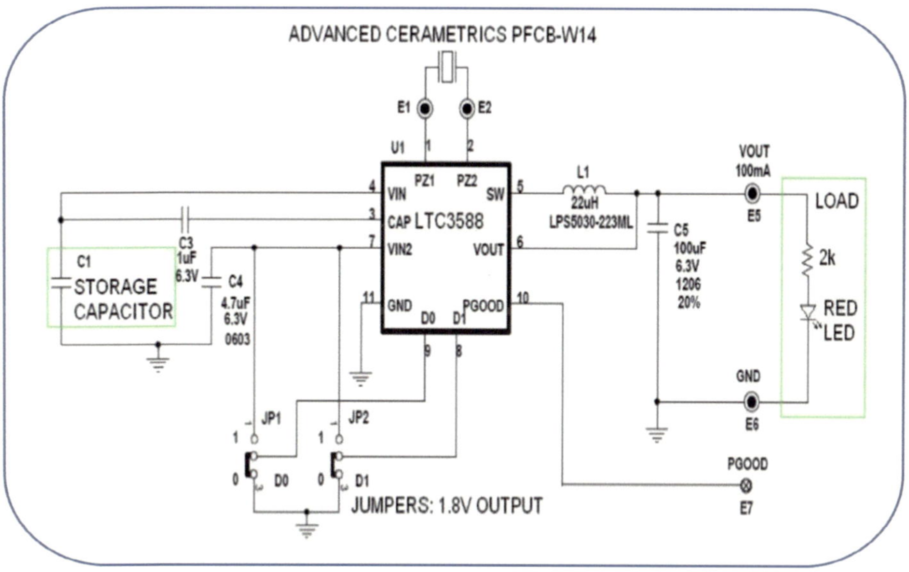

Figure 7.
Simulink model of piezoelectric-driven supercapacitor for charging and discharging.

controllable power supply network between the power transducer and the power storage element to block the power supply or to adjust the AC signal in the case of a piezoelectric device. Operating Example LTC3588–1 The output power of the transducer must exceed the minimum power limit, which increases the specific power limit set at the input pins D0 and D1. To transmit high power, the power transducer must have a double open circuit voltage and a short-circuit current of input voltage, which is twice the required input shown in **Figure 6**. These requirements must be kept to a minimum uninterruptible power supply capacity.

The piezoelectric generator, which is the constant current charge level of the supercapacitor, can be obtained by exposing the working cycle of the buck regulator using software used for pulse width modulation. The test results confirm the electrical circuit model of the piezoelectric generator, the presence of a constant current charged supercapacitor, and the simple control of the circuit designed as shown in **Figures 8** and **9**.

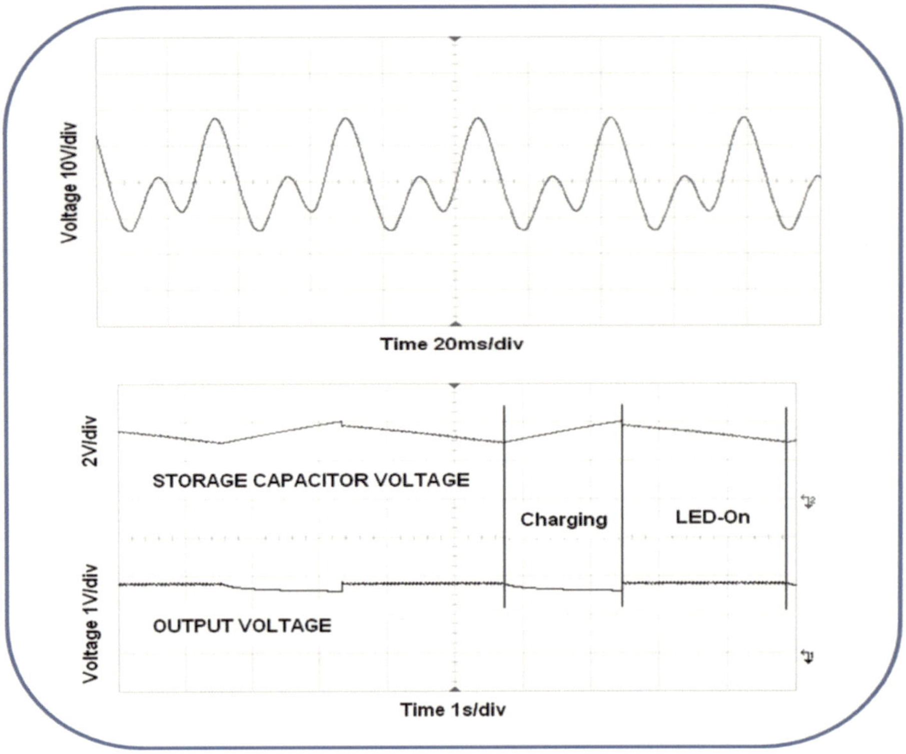

Figure 8.
Piezoelectric driven supercapacitor charging and discharging characteristics.

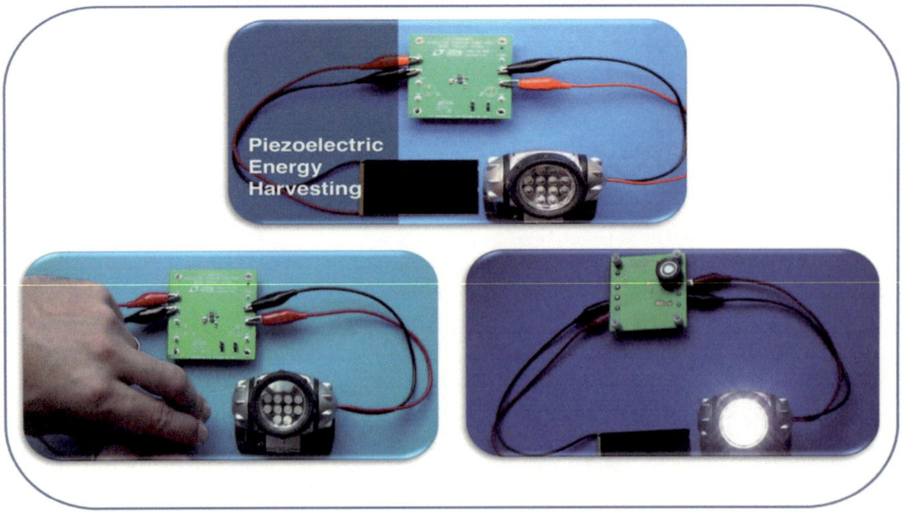

Figure 9.
Hardware implementation of piezoelectric driven charging and discharging supercapacitor for LED flash application.

5. Wearable electronics

Wearable electronic devices represent a paradigm shift in consumer electronics, body sensing, synthetic skins and wearable devices. Since all of these electronics devices require electricity to operate, portable electrical systems are an integral part of portable devices. In fact, the electrodes and other components in these electronics devices must be flexible and comfortable for the user. Presented here is a critical review of devices designed for power conversion and storage applications for use on portable devices. The main focus is on the development of solar cells, triboelectric generators, piezoelectric generators, Li-ion batteries and supercapacitors for the wearable device applications shown in **Figure 10**. These devices must be attached to the fabric., Interaction takes place. Limited to devices made with fiber and ribbon. Some major challenges and future guidelines will also be followed [24–27].

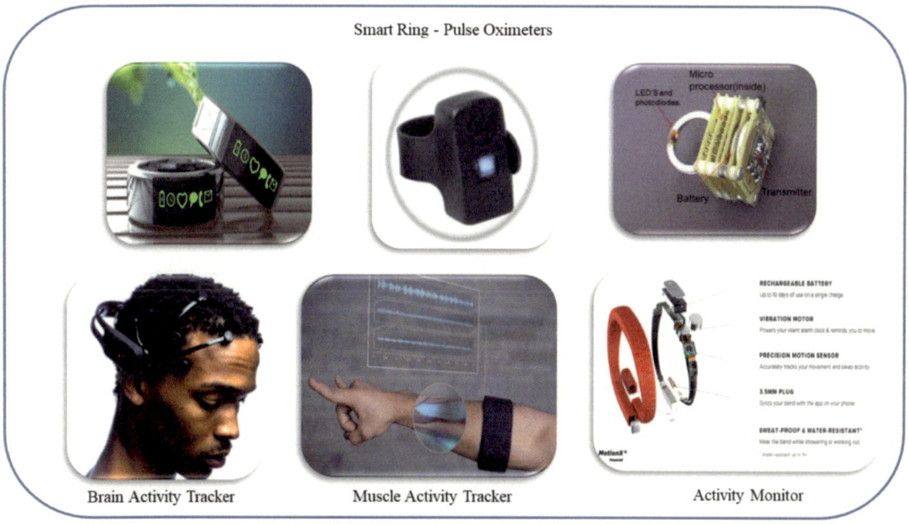

Figure 10.
Application of wearable electronics.

6. Flexible electronics

Solid state flexible supercapacitors have many applications that are flexible and wearable for current and future generations. Flexible supercapacitors can be easily connected to wearable clothing and serve as a power supply for various electronic devices such as mobile phones. The energy generated by piezoelectric generators can be stored in a large storage device and used to charge cell phones. An example is a supercapacitor-powered t-shirt called a "sound charge" that can generate electricity under the pressure of sound waves. The T-shirt tested at the Glastonbury event produced enough power to recharge two basic phones over the weekend. Activated carbon screen-printed, woven and woven with carbon fiber-based supercapacitor electrodes, fitted with a long-sleeved T-shirt, as shown in **Figure 11** [28–32].

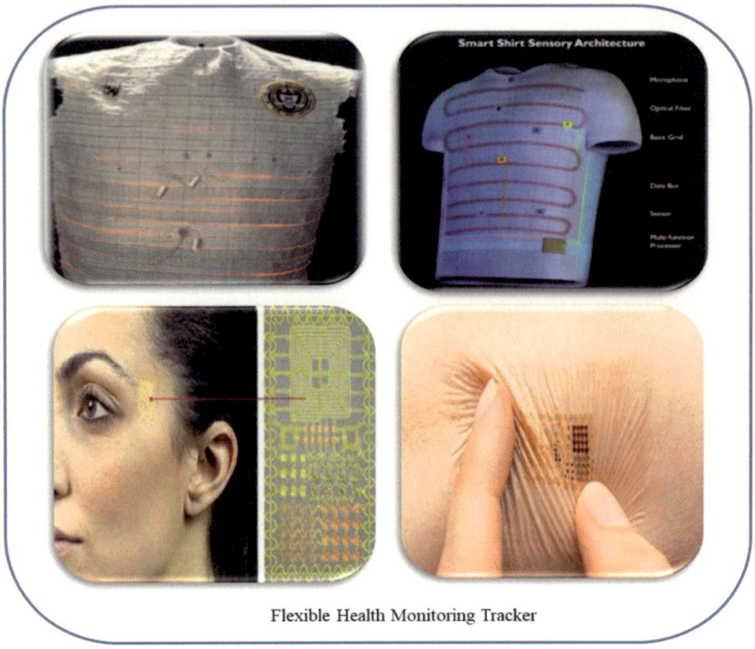

Figure 11.
Application of flexible electronics.

7. Implantable healthcare

Piezoelectric charged supercapacitors are widely used in many fitted healthcare systems where microwatts up to milliwatts are required. These supercapacitors are used for cardiac pacemakers, insulin pumps, and health care systems. Continuous glucose monitoring (CGM) systems monitor glucose levels throughout the day. CGM users inserted a small sensor wire under their skin using an automated device. This attachment has a CGM sensor housing so that the sensor can measure glucose readings in the fluid day and night. A small, reusable transmitter leads to the sensor and sends real-time readings to the receiver wirelessly, allowing the user to view the data. For some systems, a smart compatible device with a CGM application acts as a

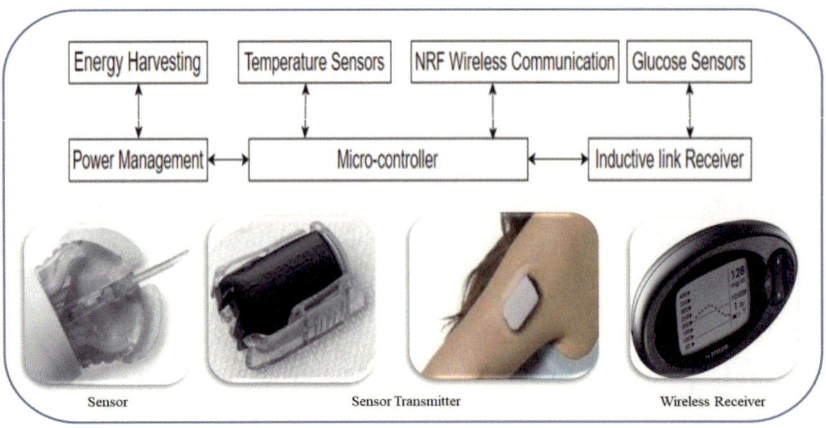

Figure 12.
Continuous glucose monitoring (CGM) sensor, receiver and transmitter.

display device. A convenient receiver or smart device reflects current glucose levels, as well as historical trends in levels. The compatible CGM receiver and/or smart device can be set to send alerts to the user when the glucose limit is reached [33–39].

With the advancement of technology in the wireless network and microelectromechanical system, smart sensors designed to be set up in remote locations, such as pull-sensing health sensors and medical sensors implanted in the human body, have lost a CGM (continuous glucose monitor) device. It provides "real time" glucose reading and trends in glucose levels. The glucose level reads under the skin every 1–5 minutes (10–15 min delay). This suggests that high and low glucose monitors turn on alarms and inform diabetes management practice. Finding a battery replacing sensor is expensive and costly every time. In embedded cases, access is impossible and devastating. With the advent of energy-efficient harvesting technology, the lifespan of those sensors can be significantly extended or replaced by their own batteries, as shown in **Figures 12, 13** and **14** [40–45].

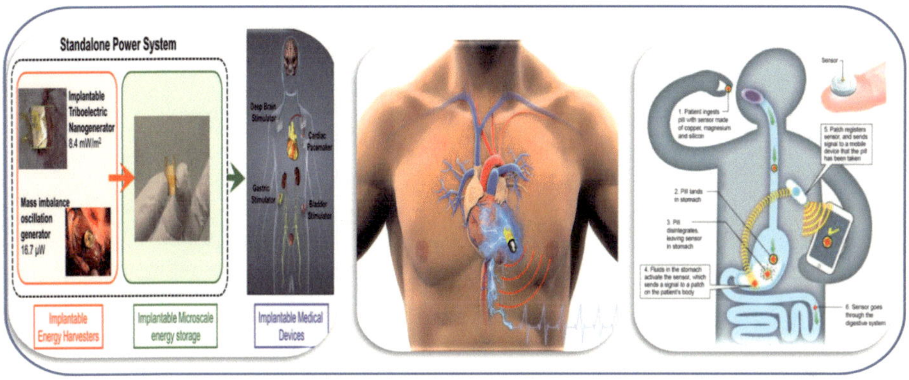

Figure 13.
Bio-medical sensor application for supercapacitor.

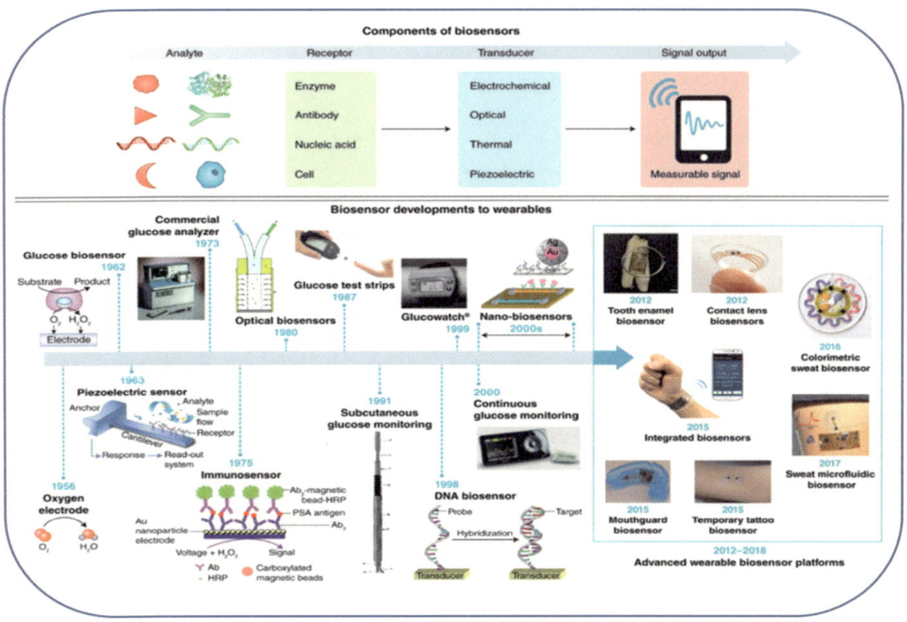

Figure 14.
Application of portable electronics and medical equipment.

8. Portable electronics

Nowadays, we cannot imagine a world without portable electronics like smartphones, smartwatch, laptops, cameras, and much more, making our daily life more modern to accomplish various new tasks. However, these smart electronics need power with energy-saving systems. Supercapacitors play an important role as an energy storage system as well as batteries. Hybrid devices with battery-supercapacitor hybrids are the best choices for current and future electronic power supply devices. This combination enables lightweight battery devices to be integrated into small portable electronic gadgets such as the clock, sensors, cell phone and headset shown in **Figure 15**.

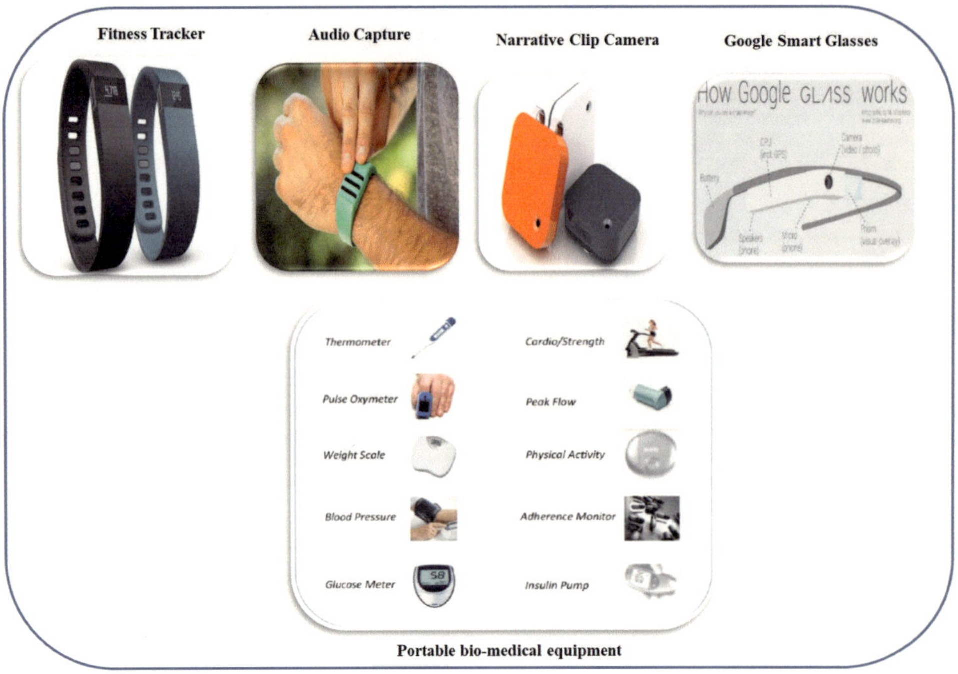

Figure 15.
Application of Portable Electronics.

9. Summary and outlook

Supercapacitors are unique energy storage systems between batteries and electrolytic capacitors based on their electrical performance. These can supply power as capacitors and bring the energy density closer to the battery. Research is not limited to the development of electrode materials and electrolytes and other materials. Many new promising supercapacitor devices are also designed for portable and wearable electronics. Made with the best development sandwiches, planers, wires, fibers, cables and more wearable and flexible supercapacitors. Research has been advanced on supercapacitors such as piezoelectric, shape-memory, thermal management systems and to extend working applications. In flexible electronics. Therefore, supercapacitors have evolved from composite applications to energy storage systems to portable and wearable electronics, smart clothing, automotive, energy-saving systems, implantable medical devices, and emerging technologies such as military and aerospace craft applications.

Author details

A. Bharathi Sankar Ammaiyappan[1,2]* and Seyezhai Ramalingam[1,2]

1 School of Electronics Engineering, Vellore Institute of Technology (VIT) Chennai Campus, Chennai, India

2 Department of EEE, Renewable Energy Conversion Lab, Sri Sivasubramaniya Nadar College of Engineering, Chennai, India

*Address all correspondence to: bharathisankar.1987@gmail.com

References

[1] Hofmann, H.F., Ottman, G.K. and Lesieutre, G.A. 2003. "Optimized Pieoelectric Energy Harvesting Circuit Using Step-down Converter in Discontinuous Conduction Mode," IEEE Trans. Power Electron., 18:696-703.

[2] Le, T.T., Han, J., Jouanne, A.V., Mayaram, K. and Fiez, T.S. 2006. "Piezoelectric Micro-power Generation Interface Circuits," IEEE J. Solid-State Circuits, 41:1411-1420.

[3] Ottman, G.K., Hofmann, H.F., Bhatt, A.C. and Lesieutre, G.A. 2002. "Adaptive Piezoelectric Energy Harvesting Circuit for Wireless Remote Power Supply," IEEE Trans. Power Electron., 17:669-676.

[4] Simjee, F.I. and Chou, P.H. 2008. "Efficient Charging of Supercapacitors for Extended Lifetime of Wireless Sensor Nodes," IEEE Trans. Power Electron., 23:1526-1536.

[5] Ramadass, Y.K.; Chandrakasan, A.P 2011 "A battery-less thermoelectric energy harvesting interface circuit with 35 mV startup voltage" IEEE J. Solid-State Circuits, 46, 333-341.

[6] Granstrom, J.; Feenstra, J.; Sodano, H.A.; Farinholt, K. "Energy harvesting from a backpack instrumented with piezoelectric shoulder straps" Smart Mater. Struct. 16, 1810-1820.

[7] George, H.B. 2010 "True Grid Independence: Robust Energy Harvesting System for Wireless Sensors Uses Piezoelectric Energy Harvesting Power Supply and LiPoly Batteries with Shunt Charger.LT J. Analog Innovation. 36-38.

[8] Rocha. J.G, Gonçalves L. M, Rocha.P. F, Silva. M. P., And Lanceros-Méndez. S. 2010 "Energy Harvesting From Piezoelectric Materials Fully Integrated In Footwear"-IEEE Transactions On Industrial Electronics, Vol. 57, No. 3.

[9] Faruk Yildiz Sam 2013 "Energy Harvesting From Passive Human Power "Houston State University International Journal of Innovative Research in Science, Engineering and Technology Vol. 2, Issue 7.

[10] Cerovsky, Z. and Mindl, P. 2005. "Regenerative Braking by Electric Hybrid Vehicles Using Supercapacitor and Power Splitting Generator," In: Proceedings of European Conference on Power Electronics and Applications, Dresden, Germany.

[11] Gualous, H., Louahlia-Gualous, H., Gallay, R. and Miraoui, A. 2007. "Supercapacitor Thermal Characterization in Transient State," In: Proceedings of 42nd IAS Annual Meeting on Industry Applications, New Orleans, LA, USA, pp. 722-729.

[12] Hofmann, H.F., Ottman, G.K. and Lesieutre, G.A. 2003. "Optimized Pieoelectric Energy Harvesting Circuit Using Step-down Converter in Discontinuous Conduction Mode," IEEE Trans. Power Electron., 18:696-703.

[13] Karthaus, U. and Fischer, M. 2003. "Full Integrated Passive UHF RFID Transponder IC with 16.7-W Minimum RF Input Power," IEEE J. Solid-State Circuits, 38:1602-1608.

[14] Kasyap, A., Johnson, D., Horowitz, S., Nishida, T., Ngo, K., Sheplak, M. and Cattafesta, L. 2002. "Energy Reclamation from a Vibrating Piezoelectric Composite Beam," In: Proceedings of 9th International Congress on Sound and Vibration, Orlando, USA, Vol. 271.

[15] Le, T.T., Han, J., Jouanne, A.V., Mayaram, K. and Fiez, T.S. 2006. "Piezoelectric Micro-power Generation Interface Circuits," IEEE J. Solid-State Circuits, 41:1411-1420.

[16] Maxwell. 2005. Charging of Ultracapacitors, Datasheet, Maxwell Technologies, Inc., San Diego, CA. Maxwell. 2009. BOOSTCAP Ultracapacitors Information Sheet, Available at: http://www.maxwell.com (accessed date January, 2009).

[17] Ottman, G.K., Hofmann, H.F., Bhatt, A.C. and Lesieutre, G.A. 2002. "Adaptive Piezoelectric Energy Harvesting Circuit for Wireless Remote Power Supply," IEEE Trans. Power Electron., 17:669-676.

[18] Ramadass, Y.K. and Chandrakasan, A.P. 2009. "An Efficient Piezoelectric Energy-harvesting Interface Circuit Using a Biasflip Rectifier and Shared Inductor," In: IEEE International Solid-State Circuits Conference, San Francisco, USA, pp. 296-297.

[19] Rashid, M.H. 2003. Power Electronics: Circuits, Devices and Applications, 3rd edn, Prentice Hall, Englewood Cliffs, NJ.

[20] Simjee, F.I. and Chou, P.H. 2006. "Everlast: Long-life Supercapacitoroperated Wireless Sensor Node," In: Proceedings of International Symposium on Low Power Electronics and Design, Tegernsee, Bavaria, Germany.

[21] Simjee, F.I. and Chou, P.H. 2008. "Efficient Charging of Supercapacitors for Extended Lifetime of Wireless Sensor Nodes," IEEE Trans. Power Electron., 23:1526-1536.

[22] Sodano, H.A., Inman, D.J. and Park, G. 2004. "A Review of Power Harvesting from Vibration Using Piezoelectric Materials," The Shock and Vibration Digest, 36:197-205.

[23] Sodano, H.A., Park, G., Leo, D.J. and Inman, D.J. 2003a. "Use of Piezoelectric Energy Harvesting Devices for Charging Batteries," In: Proceedings of SPIE 10th Annual International Symposium on Smart Structures and Materials, San Diego, CA, USA, Vol. 5050.

[24] Sodano, H.A., Park, G., Leo, D.J. and Inman, D.J. 2003b. "Model of Piezoelectric Power Harvesting Beam," In: Proceeding of ASME International Mechanical Engineering Congress and Exposition, Washington, D.C., USA, Vol. 40.

[25] Tan, Y.K., Lee, J.Y. and Panda, S.K. 2008. "Maximize Piezoelectric Energy Harvesting Using Synchronous Charge Extraction Technique for Powering Autonomous Wireless Transmitter," In: IEEE International Conference Sustainable Energy Technologies, Singapore, pp. 1123-1128.

[26] Umeda, M., Nakamura, K. and Ueha, S. 1997. "Energy Storage Characteristics of a Piezo-generator Using Impact Induced Vibration," Jap. J. Appl. Phys., 36:3146-3151.

[27] S.A. Haque et al. Review of cyber-physical system in healthcare. International Journal of Distributed Sensor Networks, 2014, 2014.

[28] Aragues et al. Trends and challenges of the emerging technologies toward interoperability and standardization in e-health communications. IEEE Communications Magazine, 2011.

[29] P. King et al. The uk prospective diabetes study (ukpds): clinical and therapeutic implications for type 2 diabetes. British Journal of Clinical Pharmacology, 1999.

[30] Murakami et al. A continuous glucose monitoring system in critical cardiac patients in the intensive care unit. In 2006 Computers in Cardiology, pages 233-236. IEEE, 2006.

[31] M. Ali et al. A bluetooth low energy implantable glucose monitoring system. In EuMC 2011, pages 1265-1268. IEEE, 2011.

[32] J. Lucisano et al. Glucose monitoring in individuals with diabetes using a long-term implanted sensor/telemetry system and model. IEEE Transactions on Biomedical Engineering, 2016.

[33] KAU. Menon et al. A survey on non-invasive blood glucose monitoring using nir. In ICCSP 2013, pages 1069-1072. IEEE, 2013.

[34] MUH. Al Rasyid et al. Implementation of blood glucose levels monitoring system based on wireless body area network. In Consumer Electronics-Taiwan (ICCE-TW), 2016 IEEE International Conference on, pages 1-2. IEEE, 2016.

[35] N. Wang and G. Kang. A monitoring system for type 2 diabetes mellitus. In Healthcom 2012, pages 62-67. IEEE, 2012.

[36] TN. Gia et al. Iot-based fall detection system with energy efficient sensor nodes. In NORCAS 2016, pages 1-6. IEEE, 2016.

[37] S. Sudevalayam and P. Kulkarni. Energy harvesting sensor nodes: Survey and implications. IEEE Communications Surveys Tutorials, 2011.

[38] M. Taghadosi et al., L. Albasha, N. Qaddoumi, and M. Ali. Miniaturised printed elliptical nested fractal multiband antenna for energy harvesting applications. IET Microwaves, Antennas Propagation, 2015.

[39] V. Jelicic et al. Analytic comparison of wake-up receivers for wsns and benefits over the wake-on radio scheme. In PM2HW2N '12, pages 99-106. ACM, 2012.

[40] L. Gu et al. Radio-triggered wake-up capability for sensor networks. In RTAS 2004, pages 27-36, 2004.

[41] Kuan-Yu Lin, T. K. K. Tsang, M. Sawan, and M. N. El-Gamal. Radio-triggered solar and rf power scavenging and management for ultra low power wireless medical applications. In 2006 IEEE International Symposium on Circuits and Systems, pages 4 pp.–5731, May 2006.

[42] S. F. Al-Sarawi. Low power schmitt trigger circuit. Electronics Letters, 38(18):1009-1010, Aug 2002. 18. M. Ali. Low Power Wireless Subcutaneous Transmitter. PhD thesis, 2010.

[43] International Commission on Non Ionizing Radiation Protection. Icnirp guidelines for limiting exposure to time varying electric, magnetic and electromagnetic fields (up to 300 ghz). Health Physics, 1998.

[44] Blood glucose monitoring. Diabetes Australia, https://www.diabetesaustralia.com.au/blood-glucose-monitoring [accessed 2016-12-22].

[45] Blood Sugar Level Ranges. Diabetes.co.uk, https://www.diabetes.co.uk/diabetescare/blood – sugar – level – ranges. html [accessed 2016 – 12 – 22].